Manufacturing for the 21st Century 2015

Topical Conference at the 2015 AIChE Spring Meeting and 11th Global Congress on Process Safety

Austin, Texas, USA
26-30 April 2015

ISBN: 978-1-5108-0691-7

Printed from e-media with permission by:

Curran Associates, Inc.
57 Morehouse Lane
Red Hook, NY 12571

Some format issues inherent in the e-media version may also appear in this print version.

Printed by Curran Associates, Inc. (2015)

For permission requests, please contact AIChE
at the address below.

AIChE
3 Park Avenue
New York, NY 10016-5991

Phone: (203) 702-7660
Fax: (203) 775-5177

www.aiche.org

Additional copies of this publication are available from:

Curran Associates, Inc.
57 Morehouse Lane
Red Hook, NY 12571 USA
Phone: 845-758-0400
Fax: 845-758-2634
Email: curran@proceedings.com
Web: www.proceedings.com

TABLE OF CONTENTS

409505 Chemical Engineering: Is This the Golden Age?

Abstract:

Schlanger will offer his insights on the current opportunities in the chemical process industries in his talk entitled "Chemical Engineering: Is This the Golden Age?" on Monday morning, April 27. At the conclusion of his lecture, Schlanger will receive the AGILE Award in recognition of his leadership at CEVA Group, a $9 billion logistics supplier to industry worldwide; LyondellBasell, a major player in the plastics, chemicals, and refining industries; Arco Chemical; and at the other companies where he has held leadership positions.

Importance of Modeling and Simulation in Framing a Process Development Career

Joseph B. Powell, Ph.D.
Shell Chief Scientist – Chemical Engineering

Overview

Process development is a key skill for industry, and the ability to develop and use economic, process, and specific technology component models is a key skill for execution of this role.

At the entry of the R&D portfolio funnel, new ideas and opportunities must be constantly screened and evaluated to select best options to purse. Screening is best accomplished via developing and comparing economic models of available options. In the absence of early information or data or performance, one can assume best possible yields and conversions, to determine which R&D options present the greatest opportunity if successfully developed. As the project progresses, the economic models are further refined to more accurately reflect required capital costs and actual yields, and to include full discounted cash flow assessments, earning power, and net present value for those projects which do progress to full development and deployment. The level of detail increases to allow a final investment decision to be made. Experienced process development engineers provide exceptional value to their organizations via the ability to provide preliminary cost estimates, odds for success, and identification of potential issues or show stoppers in early idea assessment, where limited information is available.

In Exploration and Production, modeling of enhanced oil recovery projects is the domain of supercomputers. However, a simplified model can be instructive in understanding impacts of the level of interfacial tension reduction, well spacings and capillary number, and mobility ratio of sweep efficiency, for use in process optimization. Reservoir heterogeneity often defeats the attempts of rigorous modeling via supercomputer, but does point to the fact that better characterization is needed to improve the accuracy of prediction.

Fixed-bed adsorption is another example where complex models resulting in stiff numerical equations and excessive computation and analysis time can be used, and might be the initial method of choice. However, responses typically fall into self sharpening or dispersive front behavior [1,2], such that analytical solutions for either of these limits may provide more than enough basis for commercial design, relative to the accuracy of system parameters such as mass transfer or dispersion coefficients. In the end, engineering safety factors in design may likely eliminate the need for rigorous calculation, if transport parameters were known with the accuracy needed to enable a rigorous simulation. Considerable time may be saved by recognizing these limits, and using models that are fit-for-purpose.

Pressure drop through a bed of compressible solids is another example where modeling enabled the use of model fluids for experimentation, and provided a more reliable result than direct pilot testing. Ability to develop and apply a model (modified Ergun equation) with parameters derived from small scale laboratory testing, enabled prediction of scale effects as a function of applied stress, which otherwise could not be seen without full scale testing of deployment concepts. Catastrophic failure upon scale-up from pilot to commercial was thus be averted.

Other examples include modeling of distillation via reflux experiments to obtain separation factors for use in design for real, complex mixtures, development of a commercial reactor model (hydroformylation) used to control reactor temperature profiles to avoid catastrophic loss of catalyst to extraction, development of a liquid extractor model to show the basis for counter intuitive response needed to prevent loss of interface level (via impact on slip velocity), development of effectiveness factor models to show the dramatic impact of water as solvent as a function of reactant concentration, in limiting hydrogen solubility, leading to starvation and catalyst deactivation at high volumetric reaction rates. Additional examples include adaptation of kinetics to enable substitution of a lower stench, lower volatility reaction promoter to enable low pressure reaction experiments to be conducted safely in glassware for modified catalyst development; modeling of heat losses in small scale lab reactor to project harmful runway if translated to more adiabatic, large scale commercial units; understanding that entrainment at the top of a reactor scales with linear velocity whereas reactor conversion scales with residence time, to enable high velocity testing for entrainment risk via transient operation of pilot equipment; scaling down reaction kinetics to enable poorly stirred multi throughput screening reactors to be effective in diagnosing differences in catalyst performance (vs. study of gas-liquid mass transfer), development of a transient CSTR model to enable manipulation of long-term catalyst life test units to test for catalyst activity above statistical noise level of steady state response.

Key aspects of many of these model development examples is the ability to develop fit-for-purpose, simplified or limiting-form models of key steps or principles, to generate results which are intuitively understandable and may be rapidly used to test alternate scenarios, vs. attempting rigorous modeling of all transport, rate, and equilibria, in a more complex model. The latter can allow scenarios and cases to be generated and may be warranted for some design or optimization problems, but may require considerable time and resource to develop, and underlying cause-and-effect relationships may not be readily evident. Where undertaken, results of rigorous and more comprehensive modeling should be tested via the logic and understanding of a simpler first principles model, to avoid risk of gross parameter error.

Effective use of modeling is a short cut to higher productivity in problem solving, while also providing opportunity for continued professional learning. Ability to reduce a problem to a first-principles model and provide limiting form solutions is an effective way to rapidly screen options, avoid expensive "black box testing" at larger scale, and insure accuracy of final designs.

References

1. Vermeulen, T., M. D. LeVan, N. K. Hiester, and G. Klein, *Adsorption and Ion Exchange,* Section 16 in "**Perry's Chemical Engineers Handbook**", sixth ed., R.H. Perry, D.W. Green, and J.O. Maloney, eds. McGraw-Hill, New York (1984).
2. Rhee, H.-K., B. Bodin, N. Amundson, *A study of the shock layer in equilibrium exchange systems,* **Chem. Eng. Sci**., **26**, 1571–1580 (1971).

Classical PID control in presence of missing data using compressed sensing techniques

Satheesh K. Perepu and Arun K. Tangirala*
Dept. of Chemical Engineering, IIT Madras, Chennai 600036, India
Email:satheesh841@gmail.com and arunkt@iitm.ac.in

***Abstract*—Most of the controllers used in process industries are proportional, integral and derivative (PID) controllers. Missing data is common in these industries due to many reasons such as sensor malfunctioning, data transmission loss *etc*. The performance of PID control scheme will degrade if operated in presence of missing data. Hence in this paper a method based on compressed sensing techniques is proposed for online reconstruction of missing data and performance is shown with an application to process control. The proposed algorithm is demonstrated on three simulated studies and results show that the proposed algorithm gives comparable results with classical PID control scheme with no missing data.**

***Keywords*—process control;compressed sensing;random sampling;MAR;PI control;**

I. Introduction

Maintaining the process variable at a specified level (set-point) is one of the important task in all process industries. This task can be challenging because the controller has to achieve this in presence of external disturbances. These controllers can be of basic type such as PID or of advance type such as model predictive controllers (MPC). Almost 95% of closed loop controllers used in industries are of PID type [18]. The PID controllers are most popular in industries because of their ease of maintenance and tuning. There are some hundreds of tuning rules available in literature to tune PID controllers [13]. The tuning rule used in the paper is Ziegler-Nichols open loop tuning rule which is proposed by Ziegler [19]. He developed this tuning rules based on peak sensitivity minimisation.

An important assumption made in all classical PID control schemes is that sensor measurements are available without fail at each sampling instant. However, it is not uncommon to encounter situations involving sensor and/or communication failures leading to the case of *missing samples*. Clearly such situations present several challenges to control and other operations such as monitoring and identification. In such situations a process model based on first principles will be handy. Although they provide accurate description of the process, they have two primary disadvantages (i) they require full knowledge of parameters affecting the system which is difficult to obtain (ii) solving ordinary differential equations which can be solved using numerical techniques. To address these difficulties data driven models have been developed [11], which simplify the modelling by implementing simple heuristic relationships between input and output of the system. At the same time, these data-driven models also have disadvantages: process dynamics are not fully captured by data-driven models; in addition, different data-driven models are based on different assumptions, and there is a lack of clear understanding on which assumption agrees with physical process better. Hence a method which estimates missing data online and doesn't require an explicit model is needed.

Developing methods for missing data reconstruction has been a hot topic of research for over a century now [15]. Several researchers have addressed the problem of missing samples in different applications, especially in identification, using some interesting statistical and optimal estimation approaches. However these techniques are computationally expensive restricting them to offline applications [1, 12]. To the best knowledge of authors, the problem of online missing data recovery, especially in classical control, has not been sufficiently addressed. Hence there is a need to develop a method which estimates missing data online. The main requirement of such a method is that the time needed for computing missing data should be less than that of the sampling interval chosen for the process.

In this work, a method directed at online missing data recovery for control using ideas of compressed sensing (CS) [4] is proposed. The goal in compressed sensing is to recover $\mathbf{x}$ by first estimating the sparse representation $\mathbf{s}$ given $\mathbf{A}$ or simultaneously along with $\mathbf{A}$ when it is unknown [14]. In general there will be many features in a signal like steady state, oscillatory, transients etc. Hence to find a single basis in which the entire signal has same degree of sparsity (same number of non-zeros) is difficult. Hence an idea of constructing composite dictionary [5] is used in the paper. The proposed method is discussed in §III. The two prime features of the proposed method are that, (i) it is computationally efficient and (ii) does not involve any *explicit* process model.

The rest of the paper is organized as follows. The necessary essentials of compressed sensing are reviewed in Section II. Section III presents the proposed algorithm for estimation of missing data online. Results from simulation studies are presented in Section IV. The paper closes with a few concluding remarks in Section V.

II. Foundations of Compressed Sensing

The central concept in compressed sensing (CS) is that of *sparse representation*, which refers to a minimal set of non-zero coefficients that represent the signal in some basis space. One of the main problems of interest in the compressed sensing literature is to recover N observations of a signal $z[k]$ from its M random measurements ($M \ll N$), call it $s[k]$, under the premise that $z[k]$ is K-sparse (number of non-zeros) ($K < M$) in some basis space [4]. The measurement vector $s[k]$ can be in the same domain as that of $z[k]$, for e.g., randomly subsampled

$z[k]$, or in some other transformed domain. We shall very briefly review the basic mathematics underlying these ideas below.

$$\mathbf{s} = \Pi\mathbf{z}, \quad \text{and} \quad \mathbf{z} = \mathbf{B}\mathbf{x} \tag{1}$$

$$\Longrightarrow \mathbf{s} = \mathbf{A}\mathbf{x} \tag{2}$$

where Π is $M \times N$ measurement matrix (when s is subsampled $\mathbf{z}$, it consists of ones and zeros), $\mathbf{x} \in \mathbb{R}^{N\times 1}$ is the sparse representation of $\mathbf{z}$ in some *sparsifying* basis space defined by the matrix $\mathbf{B}$. In (2), the matrix $\mathbf{A} = \Pi\mathbf{B}$ is said to be the *dictionary* for s, which may be known a priori (fixed) or unknown (adaptive) [14]. The goal in compressed sensing is to recover $\mathbf{z}$ by first estimating the sparse representation $\mathbf{x}$ given $\mathbf{A}$ or simultaneously along with $\mathbf{A}$ when it is unknown .

An N-long vector $\mathbf{x}$ is said to be sparse if the number of non-zero elements in $\mathbf{x}$ is less than the number of zero-valued elements. There exist several measures of sparsity. See [10] for a comprehensive study of different sparsity measures. The zero-norm of a vector, given by

$$||\mathbf{x}||_0 = \dim\{i|x_i \neq 0\} \tag{3}$$

is frequently used in the CS literature as a measure of sparsity.

If the dictionary $\mathbf{A}$ is known a priori, the sparse representation $\mathbf{x}$ is recovered by solving the optimization problem

$$\min_{\mathbf{x}} ||\mathbf{x}||_0 \quad \text{subject to} \quad \mathbf{s} = \mathbf{A}\mathbf{x} \tag{4}$$

However, the problem is non-convex and NP-hard. A widely used alternative is to replace with a convex 1-norm optimization to solve (2), as put forth by [4]:

$$\min_{\mathbf{x}} ||\mathbf{x}||_1 \quad \text{subject to} \quad \mathbf{s} = \mathbf{A}\mathbf{x} \tag{5}$$

Candes et. al. [2] proved that the zero-norm solution equals 1-norm solution as long as $\mathbf{A}$ satisfies what is known as the Restrictive Isometric Property (RIP), given by

$$(1-\delta)\,||\mathbf{x}||_2^2 \leq ||\mathbf{A}\mathbf{x}||_2^2 \leq (1+\delta)\,||\mathbf{x}||_2^2 \tag{6}$$

It is difficult to verify RIP property in practice. Hence, Haupt et. al. [9] proved that the Toeplitz matrices (quasi to be specific) satisfy RIP theoretically. Elad [6] replaced RIP with a weaker condition based on *mutual coherence*. The mutual coherence of matrix $\mathbf{U} = [\mathbf{u}_1 \quad \mathbf{u}_2 \quad \cdots \quad \mathbf{u}_p]$ is given by

$$\mu(\mathbf{U}) = \frac{1}{p}\max_{i\neq j} \frac{|\mathbf{u}_i^T \, \mathbf{u}_j|}{||\mathbf{u}_i||_2 \, ||\mathbf{u}_j||_2} \tag{7}$$

where $||.||_2$ is the standard 2-norm. The mutual coherence of any matrix always satisfies $1/\sqrt{p} \leq \mu(\mathbf{U}) \leq 1$; see [16]. In order to guarantee recovery of the sparse vector, the mutual coherence of the dictionary $\mathbf{A}$ should be as low as possible [6]. Tropp [17] gives a condition for the exact recovery of s in terms of the support vector of $\mathbf{A}$. The precise requirement is that at least K columns of $\mathbf{A}$ be orthogonal.

There exist two classes of algorithms to solve (5) [6]. One class is that of greedy algorithms and other comprises iterative algorithms. Greedy algorithms include orthogonal matching pursuit (OMP), block-OMP, *etc*, while the iterative algorithms include least absolute shrinkage and selection operator (LASSO), basis pursuit (BP), basis pursuit de noising (BPDN), *etc.* Among these the widely used techniques to solve (5) are the LASSO and BPDN. We choose to employ the BPDN in this work due to its robustness (to noise) property. Another advantage is that even when $\mathbf{x}$ is not sufficiently sparse, s can be approximated with a small error. In BPDN (5) is solved as

$$\min_{\mathbf{x}} ||\mathbf{x}||_1 \quad \text{such that} \quad ||\mathbf{s} - \mathbf{A}\mathbf{x}||_2 \leq \zeta \tag{8}$$

where ζ is a small non-zero value. There are several flavours of the BPDN algorithm. The spectral gradient-based BPDN (SPG-BPDN) is used in this work.

As described in §I the signal can have many features which are sparse in different basis. For example suppose some part of the signal is sparse in dictionary $\mathbf{C}$ and some part in dictionary $\mathbf{D}$. In this case matrix $\mathbf{B}$ given in (2) is constructed by concatenating two matrices $\mathbf{C}$ and $\mathbf{D}$ i.e.

$$\mathbf{B} = [\mathbf{C} \qquad \mathbf{D}] \tag{9}$$

The idea of constructing composite dictionary was proposed by [5] and they used it to reduce noise from the image. Later it was used by [8] to estimate missing data in sparse networks. In the current work we used the idea of composite dictionary to estimate missing data online and shown the performance of the method with an application to process control.

III. Proposed Algorithm

The proposed algorithm to estimate missing data online is explained as follows. Let $y[k]$ be the measured output at the kth sampling instant. Assume first N (typically very small) samples are available without any missing data. Then, the $(N+1)^{\text{th}}$ sample is predicted (without the aid of an *explicit* model) using the proposed algorithm. If the $(N+1)^{\text{th}}$ sample is missing, the predicted (estimated) value is passed on to the controller. On the other hand, when the $(N+1)^{th}$ sample is available, it is directly used to calculate the control action. It is assumed that the missing data mechanism is of *missing at random* (MAR) type.

The $(N+1)^{th}$ sample is computed using the following procedure. First, set up the equations

$$\underbrace{\begin{bmatrix} y[2] \\ \vdots \\ y[N+1] \end{bmatrix}}_{\mathbf{s}} = \underbrace{\begin{bmatrix} y[1] & \\ \vdots & \mathbf{B} \\ y[N] & \end{bmatrix}}_{\mathbf{A}} \mathbf{x} \tag{10}$$

where $\mathbf{B}$ is an $N \times N$ matrix of discrete Fourier transform (DFT) bases and $\mathbf{x}$ is the sparse representation of s in the overcomplete dictionary $\mathbf{A}$; see [7]. Note that the size of $\mathbf{A}$ is $N \times (N+1)$. The set of equations in (10) is essentially a sparse representation of the output in a *composite* basis. A useful viewpoint is that the past observations in $\mathbf{A}$ capture the dynamics of the process (as auto-regressive), while the DFT basis accounts for the steady-state/oscillatory characteristics. In (10), $y[N+1]$ and $\mathbf{x}$ are unknowns. Therefore, $\mathbf{x}$ is first estimated by leaving out $y[N+1]$ from (10),

$$\mathbf{s}(1:N-1) = \mathbf{A}(1:N-1,:)\mathbf{x} \tag{11}$$

where $\mathbf{A}(1:N-1,:)$ refers to the first $N-1$ rows of $\mathbf{A}$ and likewise for $\mathbf{s}(1:N-1)$. A standard CS algorithm such as the basis pursuit de-noising (BPDN) is used to estimate $\mathbf{x}$ in (11). Subsequently, $y[N+1]$ is estimated (predicted) as

$$\hat{y}[N+1] = \mathbf{A}(N,:)\mathbf{x}$$

To calculate $y[N+2]$, the output vector $\mathbf{s}$ and the dictionary $\mathbf{A}$ are updated as

$$\mathbf{s} = \begin{bmatrix} y[3] \\ \vdots \\ y[N+2] \end{bmatrix}, \qquad \mathbf{A} = \begin{bmatrix} y[2] & y[1] & \\ \vdots & \vdots & \mathbf{B} \\ y[N+1] & y[N] & \end{bmatrix}$$

and the above procedure is repeated. This is continued for subsequent time instants. In order to prevent the growth of $\mathbf{A}$ in size, the number of columns containing past observations may be restricted sufficiently. The value of N is chosen based on dynamics of the system. Systems with slow dynamics, i.e., large time constants (relative to sampling time), require large N, while faster dynamics call for smaller block sizes.

An important requirement of close loop controller should possess is that the algorithm should not destabilize the closed loop system. In the proposed algorithm the stability of the resulting closed-loop system is supported by the fact that $\mathbf{x}$ can be reconstructed uniquely with high probability, provided the size of $\mathbf{s}$ is $K \log N$, where K is number of non-zero elements in $\mathbf{x}$; see [4]. Its tough to prove convergence of the algorithm theoretically since the proposed one involves solving 1 norm minimization problem which doesn't have an analytical solution. Hence, the stability is proved in this paper using Monte-Carlo simulations and is discussed in §IV.

IV. Simulated Case Studies

The proposed algorithm is demonstrated on three simulated case studies. In the first two case studies, we consider both the servo and regulatory problems. In third and final case study, we consider only the servo problem. A PI controller is used in both studies and the data is assumed to be MAR, with 80% of missing data. The controllers are tuned using the standard Ziegler-Nichols closed loop tuning rules.

A. Case Study 1

The system and controller under consideration are

$$G_p(s) = \frac{1}{5s+1} \tag{12}$$

$$G_c(s) = \frac{s+1}{s} \tag{13}$$

Sampling time of the process is $T_s = 0.5$ sec and hence the value of N chosen as 10 samples. To make simulation realistic, Gaussian white noise is added to the data. Variance of white noise is calculated such that the value of SNR is 10. In this example, the percentage of missing data is coonsidered as 80%. The profile of process variable for both servo and regulatory cases are shown in Figures 1 and 2 respectively. In both the plots, the solid lines correspond to process variable, and the locations where samples are available are marked as per the legend. From the plots, it is evident that the controller performs well even in presence of high percentage of missing data. It took 0.07 sec to estimate the missing sample for the proposed algorithm which is negligible when compared to the sampling time chosen for the process. Thus the prime requirement of

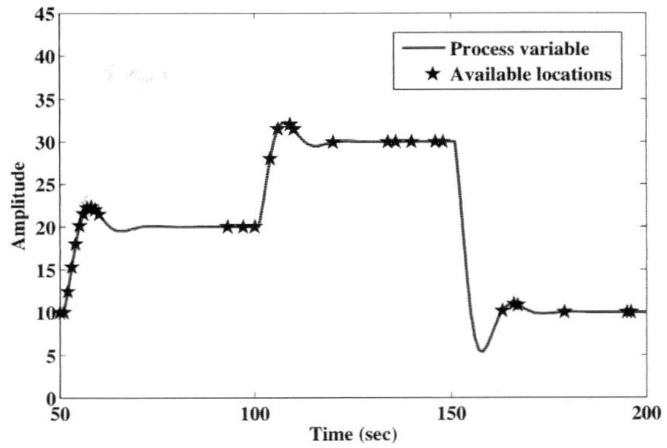

Fig. 1. First order system with PI control

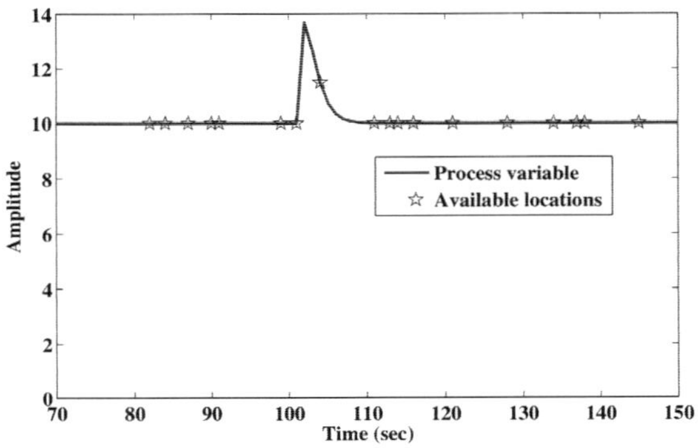

Fig. 2. First order system with regulatory control

the algorithm is satisfied. In the case of regulatory response, at least one sample should be available during the time when the effect of disturbance seen on the output of the system. In reality this seems to be a limitation for the proposed algorithm, but the probability of having no sample during the interval the effect of disturbance is seen on the output is low.

B. Case Study 2

In this case study, the system chosen is a second order system which is given by

$$G_p(s) = \frac{1}{s^2+2s+1}$$

The controller designed using Ziegler-Nichols tuning rules and is obtained as

$$G_c(s) = \frac{0.2s+1}{s}$$

The sampling time is chosen as $T_s = 0.2$ sec and the value of N chosen as 5 samples. Measurement noise is introduced such that the SNR is 10. Plots of the controlled output for the servo and regulatory cases are shown in Figure 3 and 4, respectively. The solid line in Figures 3, 4 is the output profile including both measurements and estimated data (wherever measurements are missing). The available data are marked as indicated in the legend. From the plots it is evident that the controller performs well even in presence of high percentage of missing data. The computational time required to compute the

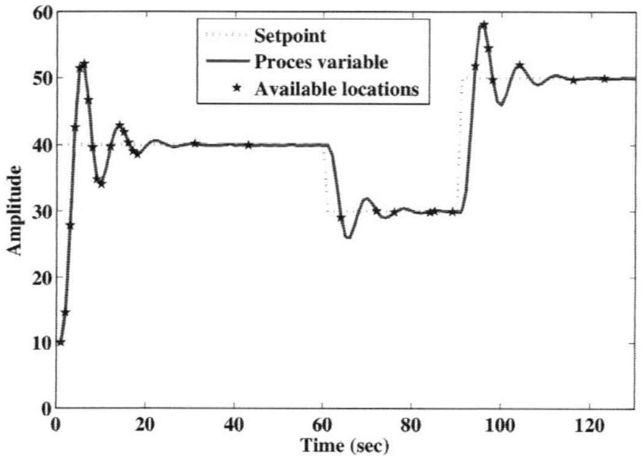

Fig. 3. Second order system with PI control

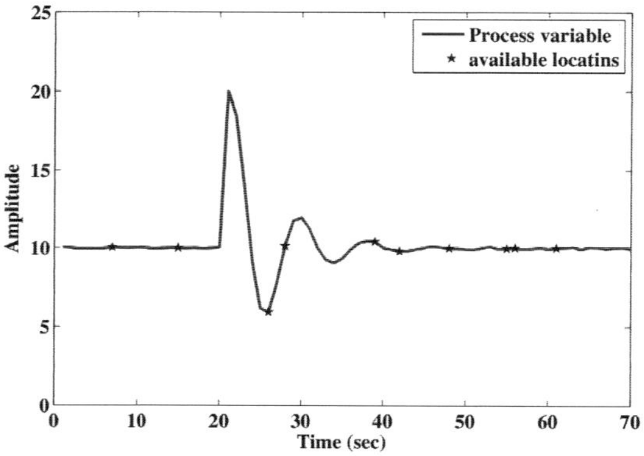

Fig. 4. Second order system with regulatory control

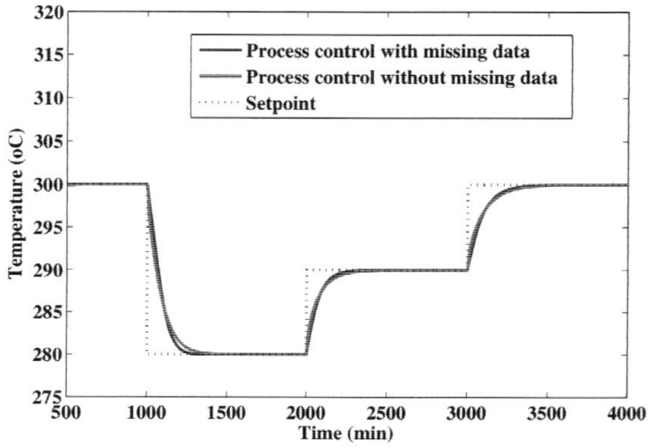

Fig. 5. CSTH with PI control

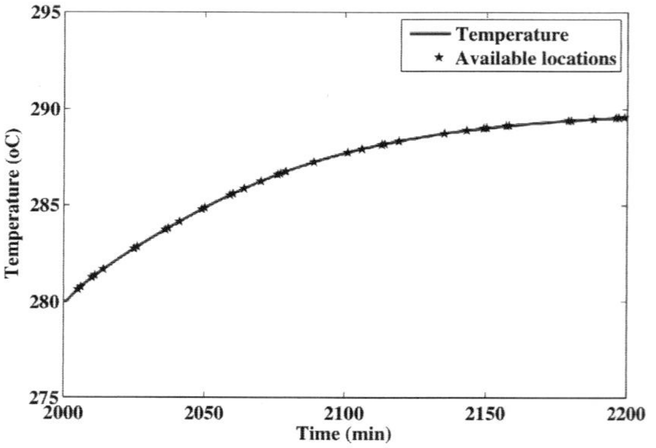

Fig. 6. Single setpoint change in temperature for CSTH

one-step ahead prediction was 0.06 sec in each case. Thus, the prime requirement of the algorithm is satisfied. As in the previous example, one needs at least one sample during the interval when the disturbance is active. Here also the probability of having no sample is during that interval is low.

C. Case Study 3

In this example a continuously-stirred tank heater (CSTH) given in [3] is simulated. Here the objective is to control the temperature of the fluid T is by manipulating the flow of liquid into jacket F_j around the heater. Like in previous case studies noise is added to the system in order to make simulation realistic. The time constant of the system is obtained as τ_p is 12 min. Hence the sampling time chosen to be $T_s = 2$ min and value of N is chosen as 20 samples. The time profile of the controlled variable resulting from the implementation of the proposed algorithm, for set-point changes, are shown in Figures 5. For better readability the plot of process variable with the locations of true measurements is shown in Figure 6. For comparison purpose, the profile of controlled variable for no missing data case is also shown in Figure 5. The controller tracks the setpoint well even in presence of high percentage of missing data. The time taken to compute each missing sample is 0.15 sec which is negligible when compared with sampling time.

The three case studies demonstrate the potential of the proposed method for control in presence of missing data, in both servo and regulatory cases.

D. A numerical study of stability of the proposed algorithm

As discussed at the end of §III stability of the algorithm is verified using Monte-Carlo simulations and is demonstrated on CSTH example described in previous case study. The experiment is repeated for 100 times and integral absolute error (IAE) value obtained after each run is plotted in Figure 7. The IAE is calculated using the formula

$$\text{IAE} = \sum_{k=0}^{q-1} |T - T_{sp}| \tag{14}$$

where T_{sp} is setpoint of the temperature and q is simulation time.

From the plot, it is evident that the IAE is almost constant in all simulations which proves that the algorithm is converging in all iterations, there by ensuring the stability of the algorithm.

V. Conclusion

In this paper a method was presented to estimate missing data online using CS techniques. In addition we shown the performance of the method with an application to classical PID control in presence of missing data. The proposed algorithm is demonstrated using three simulated examples. All the case

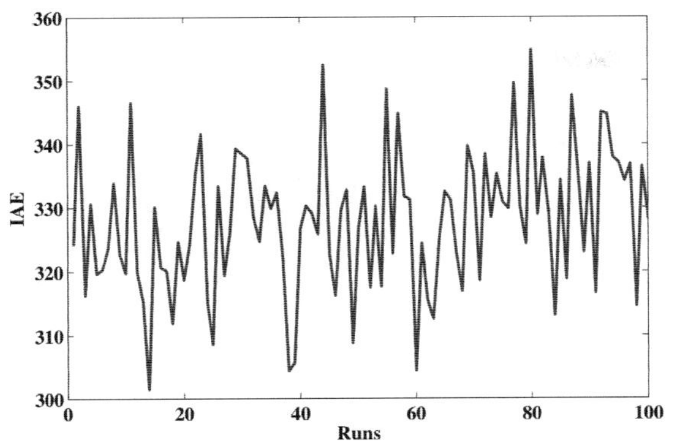

Fig. 7. Runs vs IAE

studies demonstrate the efficacy of the proposed method and shows comparable performance with full data case. At last we discussed the stability of the proposed algorithm and is proved using Monte-Carlo simulations on CSTH. Future directions include developing theoretical guarantees of the stability and extending the proposed algorithm for control of large scale systems.

REFERENCES

[1] T. Bohlin. On the problem of ambiguities in maximum likelihood identification. *Automatica*, 7(2):199–210, March 1971.

[2] Emmanuel J. Candès. The restricted isometry property and its implications for compressed sensing. *Comptes Rendus Mathematique*, 346(9–10):589 – 592, 2008.

[3] Denis Dochain. Process dynamics modeling, analysis, and simulation. *The Canadian Journal of Chemical Engineering*, 77(1), 1999.

[4] D.L. Donoho. Compressed sensing. *Information Theory, IEEE Transactions on*, 52(4):1289–1306, April 2006.

[5] M. Elad and M. Aharon. Image denoising via sparse and redundant representations over learned dictionaries. *Image Processing, IEEE Transactions on*, 15(12):3736–3745, 2006.

[6] Michael Elad. *Sparse and Redundant Representations: From Theory to Applications in Signal and Image Processing*. Springer Publishing Company, Incorporated, 1st edition, 2010.

[7] D. Guo, L. Zicheng, Q. Xiaobo, and L Huang. Sparsity-Based Online Missing Data Recovery Using Overcomplete Dictionary. *Sensors Journal, IEEE*, 12(7):2485–2495, 2012.

[8] Di Guo, Zicheng Liu, Xiaobo Qu, Lianfen Huang, Yan Yao, and Ming-Ting Sun. Sparsity-based online missing data recovery using overcomplete dictionary. *Sensors Journal, IEEE*, 12(7):2485–2495, July 2012.

[9] J. Haupt, W.U. Bajwa, G. Raz, and R. Nowak. Toeplitz compressed sensing matrices with applications to sparse channel estimation. *Information Theory, IEEE Transactions on*, 56(11):5862–5875, Nov 2010.

[10] N. Hurley and Scott Rickard. Comparing measures of sparsity. *Information Theory, IEEE Transactions on*, 55(10):4723–4741, Oct 2009.

[11] Lennart Ljung. *System Identification - A Theory for the User*. Prentice Hall International, New Jersey, USA, 1999.

[12] T Moon. The Expectation Maximization Algorithm. *IEEE Signal Processing Magazine*, pages 47–60, 1996.

[13] A. O'Dwyer. *Handbook of PI and PID Controller Tuning Rules*. Imperial College Press, 2009.

[14] Satheesh Perepu and Arun K. Tangirala. An adaptive basis estimation method for compressed sensing with applications to missing data reconstruction. In *DYCOPS 2013*, pages 190–196, Mumbai, India, 2013. IFAC.

[15] C. R Rao, T Helge, Shalabh, and H Christian. *Linear Models and Generalizations*. 1995.

[16] J. Romberg and M. Wakin. Compressed Sensing: A Tutorial. In *IEEE Statistical Signal Processing Workshop,*, 2007.

[17] J.A. Tropp. Just relax: convex programming methods for identifying sparse signals in noise. *Information Theory, IEEE Transactions on*, 52(3):1030–1051, March 2006.

[18] Jinghua Zhong. Pid controller tuning: A short tutorial. *Mechanical Engineering, Purdue University*, 2006.

[19] J. G. Ziegler and N. B. Nichols. Optimum settings for automatic controllers. *Transactions of ASME*, 64:759–768, 1942.

WATERFALL®

Stronger Than Firewalls

Evolving cyber security best practices for network perimeter protection

Michael Firstenberg, Director of Industrial Security
Waterfall Security Solutions

WATERFALL®

Cyber Perimeter - How Secure are Firewalls Really?

Attack Type	UGW	Fwall
1) Phishing / drive-by-download – victim pulls your attack through firewall		
2) Social engineering – steal a password / keystroke logger / shoulder surf		
3) Compromise domain controller – create ICS host or firewall account		
4) Attack exposed servers – SQL injection / DOS / buffer-overflowd		
5) Attack exposed clients – compromised web svrs/ file svrs / buf-overflows		
6) Session hijacking – MIM / steal HTTP cookies / command injection		
7) Piggy-back on VPN – split tunneling / malware propagation		
8) Firewall vulnerabilities – bugs / zero-days / default passwd/ design vulns		
9) Errors and omissions – bad fwall rules/configs / IT reaches through fwalls		
10) Forge an IP address – firewall rules are IP-based		

Attack Success Rate:	Impossible	Routine	Easy

Photo: Red Tiger Security

Firewall have been with us for 30 years now. The good guys and the bad guys both know how to defeat them

 2

WATERFALL®

Unidirectional Security Gateways: Server Replication

- Hardware-enforced unidirectional server replication
- Replica server contains all data and functionality of original
- External clients communicate only with replica historian
- 100% secure from online attacks from external networks
- Replicate historian servers, OPC servers, RDB servers, Modbus, etc.

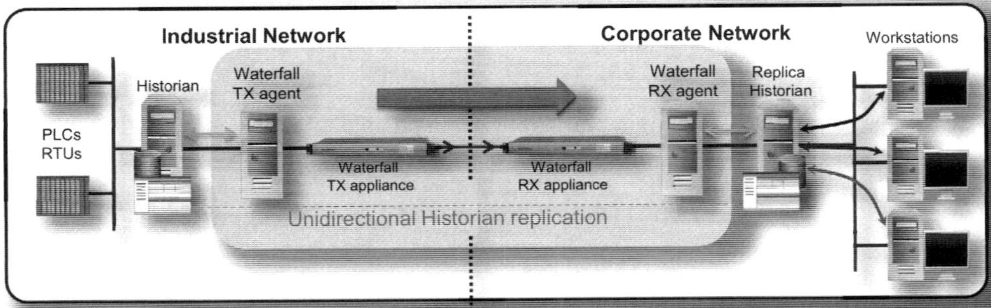

 3

WATERFALL®

Modular Hardware Package

- Can use Waterfall host modules, or customer-supplied servers
- Supports dedicated TX/RX agent hosts, or shared, or virtual
- Supports all-in-one 1u rack-mount form factor, or each module in separate cabinet, or any combination
- Supports Windows and Linux
- User serviceable, user-expandable
- Standard: 2x dual power supplies
- Standard: 1Gbps

Modular, powerful, flexible, and secure gateway hardware system

 4

WATERFALL

Waterfall FLIP™ Defeats Interactive Remote Control

- Unidirectional Gateway whose direction can be reversed:
  - Chemicals / refining / mining / pharmaceuticals: batch instructions
  - Water systems: periodic security updates & anti-virus signatures
  - Remote unstaffed sites: substations, pumping stations
- Trigger: button / key, schedule
- Stronger than firewalls, stronger than removable media

The FLIP is a Unidirectional Gateway that can "flip over"

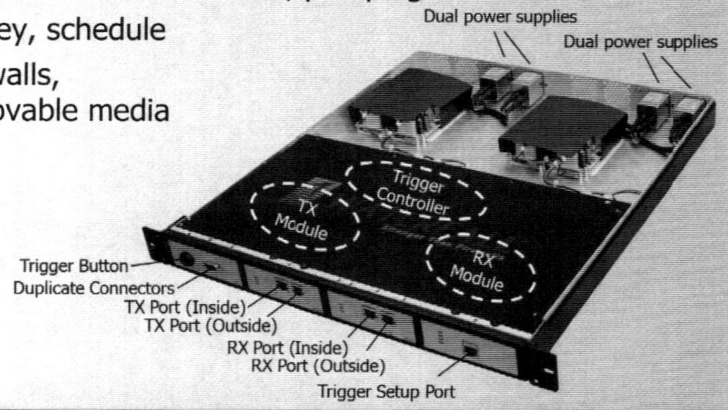

WATERFALL

FLIP: Stronger than Firewalls

- The FLIP is a Unidirectional Security Gateway – it can never be bi-directional
- The FLIP prevents interactive remote control – it can not FLIP fast enough to permit Remote Desktop or interactive SSH sessions
- Trigger mechanism cannot be subverted by network attacks
- Firewalls forward messages, the FLIP & Gateways do not
- No protocol-level attacks pass through – no fuzzing/buffer overflows. All communications sessions terminate in agent hosts.

FLIP: Stronger than firewalls

WATERFALL®

Application Data Control

- Powerful policy-driven application data control filters
  - Read and write value bounds checking, limiting
  - Limit commands, limit points & tags
- Apply policies to even proprietary, encrypted data streams
- Pass files & BLOBs to arbitrary third-party virus scanning, cryptographic signature verification & other tools

Waterfall ADC delivers on promises long made by next-gen firewalls

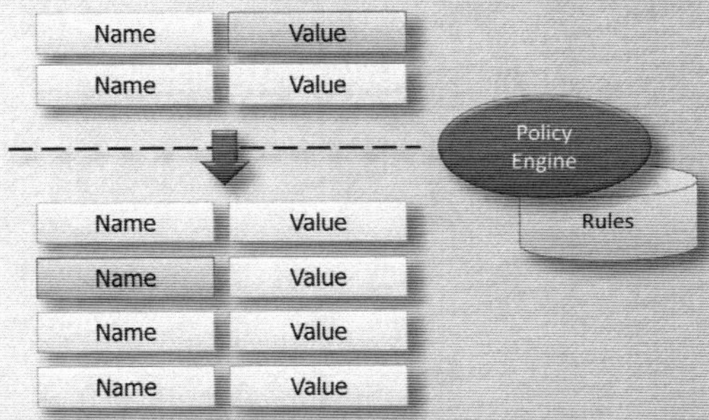

 7

WATERFALL®

Evolving Best Practices – Unidirectional Gateways

NERC
NORTH AMERICAN ELECTRIC RELIABILITY CORPORATION

NERC CIP exempts unidirectionally-protected sites **from over 35% of requirements**

ICS-CERT
INDUSTRIAL CONTROL SYSTEMS CYBER EMERGENCY RESPONSE TEAM

DHS recommends unidirectional gateways in **security assessments** (ICS CERT)

UNITED STATES NUCLEAR REGULATORY COMMISSION
NEI
NUCLEAR ENERGY INSTITUTE

NRC & NEI exempts unidirectionally-protected sites from **21 of 26 cyber-perimeter rules**

NIST
National Institute of Standards and Technology
U.S. Department of Commerce

NIST - unidirectional gateways prevent any connectivity of **traffic between domains** (800-82)

enisa
European Network and Information Security Agency

ENISA - unidirectional gateways provide **better protection than firewalls**

ISA
IEC International Electrotechnical Commission

Unidirectional gateways – **limit the propagation of malicious code** (ISA SP-99-3-3 / IEC 62443-3-3)

 8

WATERFALL®

Which of Our Networks are Expendible?

- Attacks only become more sophisticated over time
- Modern attacks routinely defeat firewalls & other security software
- Best practices are evolving – hardware-enforced Unidirectional Security Gateways are stronger than firewalls
- Absolute protection from network attacks originating on external neworks

FROST & SULLIVAN

"Waterfall's unique solutions have the potential to be the industry's next game changing standard"

So – which of our networks are expendable enough to protect with software alone?

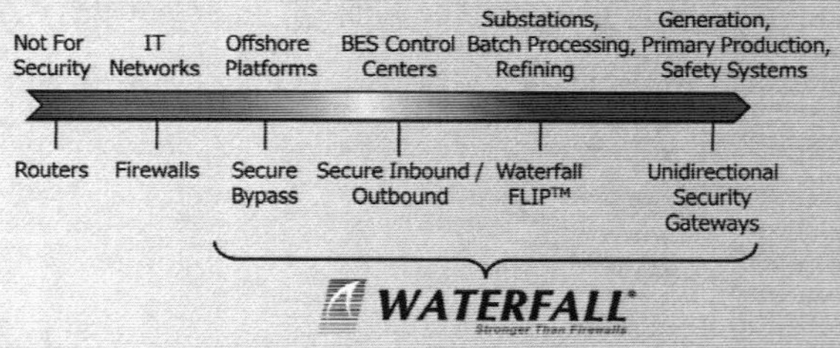

 9

398090 On Parabolic and Hyperbolic Diffusion Effects during Hybridization Kinetics with Convection in Microfluidic Chamber

Abstract:

The hybridization reaction between target molecules that is going to be studied with the probe molecules that are used for detection can be seen to be reversible reaction. Thus,

$$P + T \underset{k_r}{\overset{k_f}{\quad}} P\text{-}T \qquad (4.27)$$

where P is the probe molecules, T – target molecules and P-T the probe-target hybridized duplex that is formed on the substrate.

(4.28)

Where [P] is the concentration of the probe molecules, [T] is the concentration of the target molecules, [P-T] is the concentration of the probe-target duplex molecules, k_f is the rate constant of the forward reaction and k_r is the rate constant of the reverse reaction. The ratio of [P-T] to [P] provides a measure of the efficiency of the hybridization.

0

$[P]_0$ is the concentration of the fresh probe molecules available for tagging in the beginning of the operation. This also can be seen to be the maximum concentration of probe molecules during the reaction process. When the target molecules are assumed to be in excess the rat expression in Eq. () can be seen to obey pseudo-first order kinetics. The rate would only vary with the concentration of the probe molecules and probe-target duplex molecules and the concentration of the target molecules can be lumped along with the rate constant, k_f and can be expected to remain constant during the reaction process. The rate expression can be simplified as;

(4.29)

Assuming that at t = 0, no probe-target duplex molecules were formed, i.e., $[P\text{-}T]_0 = 0$, Eq. () can be integrated as follows;

(4.29)

Where

Integrating Eq. ();

(4.30)

or

(4.31)

(4.32)

Where K is the equilibrium rate constant given by; . Eq. () is for the ideal case. This represents the highest rate possible for hybridization. Diffusion effects can result in a lower overall rate of hybridization.

The relative effects of intrinsic hybridization reaction rate and diffusion rate were evaluated using a Damkohler number. Damkohler number was defined by Pascault et. Al. [2008] as the ratio of maximal forword reaction rate to the maximum normal diffusion rate. When Da is greater than 1 they found the process to be diffusion limited.

Sharma [2005] has shown that when the dimensions of the chamber is small and the time of transport are low damped wave diffusion effects may become important. The relaxation times of the oligonucleotides may also have to be considered. The relaxation times can be expected to increase with the increase in size of the oligonucleotide. When the relaxation times are large and/or the microfluidic chamber size is small the concentration of the probe molecules can be expected to undergo subcritical damped oscillations. The probe molecules will have to meet the target molecules and form the duplex. The duplex molecules can be attached to the microarray substrate using linker molecules. Optimal probe concentration is discussed in Sharma [2009]. Quality of surface needed is also discussed. Single stranded DNA molecule will be 12 A^0 in diameter. So the P-T duplex molecules will be 24 A^0 in diameter. Due to major and minor grooves the effective diameter will be 20 A^0. The optimal target concentration is achieved when the spacing of target molecules are at least 20 A^0 each. Further packing will cause steric hindrance.

Consider the microfluidic chamber to be made of a finite slab with thickness of 2a (Figure 4.3). At one end of the surface the probe concentration is at the maximum and the constant wall concentration boundary condition can be assumed. The other end can be assumed to be impervious to the probe concentration. As the probe molecules diffuse through the target solution simultaneous reaction and damped wave diffusion effects may become important.

Figure 4.3 Microfluidic Chamber for P-T Duplex Formation

The governing equation for simultaneous damped wave diffusion, relaxation and reaction that obeys the irreversible first order kinetics, can be written as follows;

(4.33)

Where C_p is the concentration of the probe molecules, t_r is the relaxation time of the probe molecules, D_{pt} is the binary diffusion coefficient of the probe molecules in target solution, k is the rate constant of the probe-target duplex formation assuming that the probe-target reaction is first order and irreversible and $k^* = kt_r$. The governing equation can be made dimensionless by the following substitutions;

(4.34)

The governing equation becomes;

(4.35)

The boundary conditions can be written as follows;

X = 0, u = 0 (4.36)

(4.37)

The governing equation needs to be made homogenous. The solution can be assumed to comprise of two parts, i.e., steady state part and a transient state part. Thus,

(4.38)

Pluggind Eq. (4.38) in Eq. (4,35) the governing equation can be written as two equations;

(4.39)

(4.40)

The governing equation to the steady state part is a second order differential equation with constant coefficients. The k* term will result in a integrating factor of another constant a_3. The solution to the second order differential equation with constant coefficients can be written as a sum of sinh and cosh hyperbolic functions. The integrating factor a_3 can be seen to be -1 by a $g = (1+ u^{ss})$ substitution. Thus;

(4.41)

Applying the boundary conditions to the solution a_2 can be seen to be 1 from the X = 0, u = 0 condition. Applying the impervious boundary condition at x = 2a;

(4.42)

(4.43)

(4.44)

The solution to the transient part of the dimensionless concentration can be obtained by the method of separation of variables. The dimensionless transient concentration can be expressed as a superposed product of decaying exponential and wave concentration. This can also be arrived at by multiplying Eq. () by e^{nt}. By choice of n and grouping the terms the wave concentration can be seen to be;

(4.45)

For , the wave concentration can be seen to obey;

(4.46)

The hyperbolic second order partial differential equation for wave concentration can be solved for by the method of separation of variables. Let W = g(t)f(X). Then Eq. () becomes;

(4.47)

The space domain solution can be written as follows;

(4.48)

From the boundary condition at X = 0, c_2 can be seen to be 0. From the impervious boundary condition at x = 2a;

(4.49)

(4.50)

Or, , n = 0, 1, 2, 3, 4,....... (4.51)

The time domain solution can be seen to be;

(4.52)

It was shown by Sharma [2005] that final condition in time can be applied and solutions within the frameword of second law of thermodynamics can be obtained. When t ®∞, what happens to the wave concentration ?

(4.53)

The dimensionless concentration u becomes 0 and the exponential becomes infinity. Although the product of 0 and infinity is of the indetermintate form of the fourth kind [Piskunov, 1965] it can be assumed that W will be finite or zero at infinite time and will not be infinity. Applying this condition to the time domain solution it can be seen that c_3 is 0. The generation solution for the transient concentration can be written as sum of infinite modified Fourier series and;

(4.54)

The initial condition can be used in order to obtain the c_n by invoking the principle of orthogonolity. C_n can be seen to be . When the relaxation times are large the concentration of the probe can be expected to undergo subcritical damped oscillations. This is when;

(4.55)

This can happen with more probe molecular types especially when a is small. This is the case for the microfluidic chamber in Figure 4.3. This is about 100 μm. k* also contains a contribution from relaxation time, t_r. For systems with slow reaction and higher diffusivity Eq. () will yield a lower threshold value for the relaxation time. For systems with large relaxation times the concentration profile can be written as follows;

400957 Dynamic Simulation for Poly-Ethylene Process By PC-SAFT Model

Abstract:

Current applications for polyethylene plant simulation, especially for Operator Training Simulators (OTS), safety features analysis in dynamic, as well as ethylene compressor control mechanism, require both model accuracy and simulation performance. However this cannot be achieved without proper selection of the thermodynamic model. Perturbed-Chain Statistical Associating Fluid Theory (PC-SAFT) [1] has been well established through its accuracy in modeling monomer solubility into polymer solutions, and hence placed itself as an excellent candidate for model accuracy of polyethylene plant simulation.

Questions for PC-SAFT come from the performance and reliability of the density solutions on vapor/liquid phase at extreme conditions such as high temperature and high pressure [2]. Here, we present a state of the art implementation and modification of PC-SAFT through DYNSIM, a dynamic simulation tool by Schneider Electric that achieves both model accuracy and simulation performance, with which clients can apply to execute various tasks in both steady state and dynamic analysis over the entire plant.

Detailed information for the implementation and modification of PC-SAFT would be discussed in the scopes of

- Super-linear approach when previous rigorous solution had been converged;
- Pseudo-density extrapolation near and over super-critical region

Demonstration by DYNSIM simulation with PC-SAFT at key regions of polyethylene plant would be displayed, followed by the potential upstream and chemical plants' applications.

[1] J. Gross and G. Sadowski, "Perturbed-Chain SAFT: An Equation of State Based on a Perturbation Theory for Chain Molecules", Ind. Eng. Chem. Res. 2001, 40, 1244-1260

[2] Privat R., R. Gani and J-N. Jaubert, "Are safe results obtained when the PC-SAFT equation of state is applied to ordinary pure chemicals?", Fluid Phase Equilib. 2010, 1, 76-92.

399610 Advances in Liquid Level Systems for Process Control Education and Researches

Abstract:

Process control is closely related with static and dynamic characteristics of processes. Experiments will improve the process control course. For this, small experimental systems without producing wastes have been proposed in AIChE spring meeting 2012 (Apr. 1-5, Houston, Texas, U.S.A.) by the authors. Many advances have been made since then and will be presented here.

For the purpose of process control education, liquid level systems are adequate, but most of liquid level systems available are too simple to illustrate various process static and dynamic characteristics (Astrom, K.J. and Ostberg, A.B., "A Teaching Laboratory for Process Control," IEEE Control Systems Magazine, 6, 37-42, 1986). On the other hand, the four tank system suggested by Johansson (Johansson, K.H. "The Quadruple-Tank Process: A Multivariable Laboratory Process with an Adjustable Zero," IEEE Trans. Control Systems Technology, 8, 456-465, 2000) provides a system with unstable zero, which gives a nontrivial control problem, and is found to be very useful for process control education and researches. Several research papers have also been available (for example, Mercangoz, M. and Doyle III, F.J., "Distributed Model Predictive Control of a Four-Tank System," J. Process Control, 17, 297-308, 2007). The four tank system requires four tanks, four manual valves, two actuators and two level sensors. Our previous liquid level systems have been less complex and have more functions, realizing most of process characteristics which are in process control textbook (Seborg, D.E., Edgar, T.F. and Mellichamp, D.A., Process Dynamics and Control, 2nd ed. Wiley, N.Y., 2004).

Functions of the proposed liquid level system include (1) Linear Element with 1st Order, 2nd Order, and so on, (2) Integral Element, (3) Time Delay Element, (4) Inverse Response Element (Unstable Zero Element), (5) S-shape Nonlinear Element (Mimic pH Process), (6) Input Multiplicity Element, (7) Hysteresis Element (Mimic Ignition of Exothermic Reactor), and (8) Limit Cycle Instability Element. Functions newly added in this work are (9) Unstable Element, (10) Under-damped Element, and (11) Large Negative Zero Element.

WATER INJECTION PUMP'S WASTE HEAT RECOVERY SYSTEM RETROFIT TO ENHANCE ENERGY EFFICIENCY

Zeeshan Farooq, Mahmoud Bahy Nour Eldin and Omar Rajeh, P&CSD Saudi Aramco, Dhahran

Abstract

Water Injection Pumps (WIPs) are essential to oil production, it injects water back into the reservoir to increase pressure and thereby stimulate production. The high pressure, high flow WIPs are big consumer of energy and if they are driven by Gas Turbines (GTs) the exhaust gases gives a lot of opportunity for energy integration. In this paper, Water Injection Pumps (WIPs) driven by Gas Turbines (GTs) using sweet gas as fuel is investigated for best option for waste heat recovery. We demonstrated that the proper utilization of waste heat with the available technology can lead to efficient WIPs system design that results in energy saving and energy based GHG emission reductions. Several technologies available today can retrofit old systems and design new expansions effectively to increase the overall system efficiency.

Introduction

Improving waste heat recovery systems in industrial facilities can reduce the overall demand for energy consumption. Waste heat can be utilized to reduce power import or also can even to produce power with the help of today's new technologies. In this paper we have system that consists of Water Injection Pumps (WIPs) driven by Gas Turbines (GTs) utilizing Sales Gas (sweet gas) as fuel. The GTs exhaust is utilized for Heat Recovery Steam Generators (HRSGs) to produce MP steam to be used as heating utility for Gas and Oil Separation Plants (GOSP). The GOSP has expansion plans to increase oil and gas production. The flue gas exhaust system is designed for 400^0F; this high exhaust temperature is used to enable use of back-up fuel gas (sour gas) in emergency cases when Sales Gas is not available. A new system is proposed to produce more steam by utilizing heat available in the flue gas, which can be switched back to the current system. In order to maximize the waste heat recovery efficiency the produced steam is used for Chilling in Gas Train, inlet cooling to GTs and power production via Condensing turbine. This case study demonstrate that huge savings are possible when the current design i.e. exhaust flue gas temperature of 400^0F shall be lower to exhaust temperature of 250^0F.

Process Description

The GOSP presented in this study consists of four 3-phase Separator and Gas Compression units and two Gas Trains consists of Dehydration (TEG) and NGL units. This facility also has planned incremental expansion to have one more GOSP and one more Gas Train. System can be represented pictorially as follows;

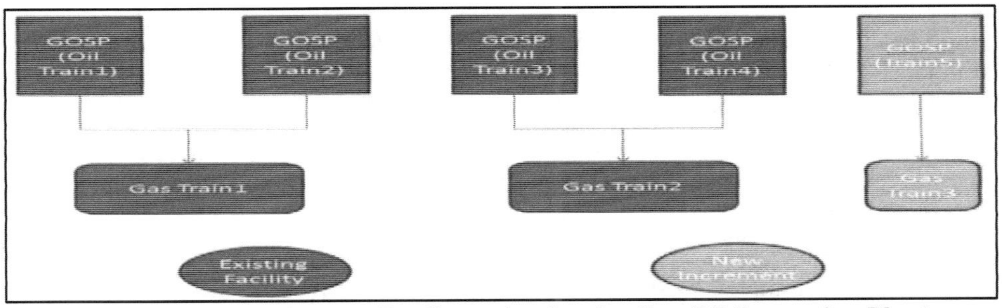

Figure 1: Process System Configuration for the Gas-Oil Separation Plant

The main process utilities are mention below and represented pictorially in Figure 2.

1. Power (Exported from outside)
2. MP Steam (Produced from HRSGs and Let down for LP steam)

3. Cooling Water (Cooling Tower)
4. Air (Fin-fans)
5. Chilling (Vapor Compression Refrigeration Cycle)

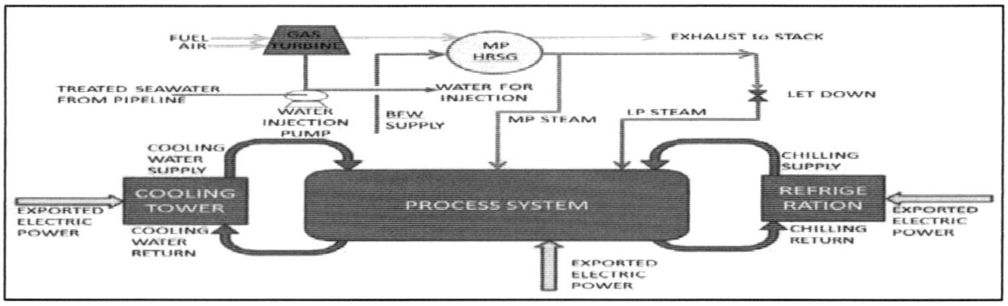

Figure 2: Utility System Configuration for the Gas-Oil Separation Plant

The system shows that the heat (steam) provided to the process is almost free (provided from HRSGs (Heat Recovery Steam Generators)) and any heat integration will results in saving steam which is the only heating utility results in no monetary savings. It is very clear from the system big picture that any steam saved won't result in appreciable monetary savings; hence no steam saving initiatives will justify investment. This lead us to the conclusion that any saved steam needs to be utilized in a better way to generate monetary savings to the facility. Moreover, it is noteworthy to pay attention that facility requirement for direct power as well as power for vapor compression cycle is large. Therefore, there is a need to explore the available technologies to utilize steam for power and chilling requirement.

Problem Statement

Given WIPs system with waste heat recovery in place and new expansion plans, identify options/scenarios for both existing and new waste heat recovery system to improve the overall system's efficiency.

Solution Approach

Solution approach adapted is to do targeting for the GT exhaust flue gases to produce HP, MP steam. Utilize excess steam (remaining after fulfilling process heating requirement) for all possible configurations (with available proven technologies) to make all possible cases to be evaluated with CHP model in terms of Net Present Value (NPV).

Water Injection Pumps (WIPs) in consideration are Gas Turbines (GTs) driven pumps. Steam generation is done through waste heat boilers, utilizing GTs exhaust to produce MP (560 mpph, 140 mpph per WIP) steam. Facility steam requirement is Medium Pressure (512 mpph) and Low Pressure (48 mpph) Steam, MP steam is let down to be utilize as LP steam through let down stations. CHP system is simple and no power can be produced with this arrangement. Present facility have 5 WIPs driven by GTs with four GTs exhaust is utilized in HRSGs to produce steam. Future facility as per planned expansion shall have seven WIPs all driven by GTs and all exhaust shall be utilized in HRSGs to produce MP steam with same arrangement results in inefficient system. Currently all steam produced from HRSGs is utilized by the facility but with the new increment there is unbalance in generation and consumption. The MP steam demand increases from 560 to 870 mpph and LP steam demand increases from 48 to 60 mpph.

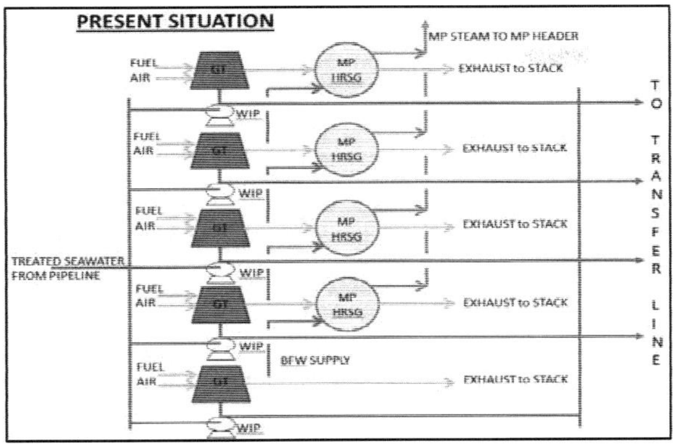

Figure 3: Present Steam Generation Arrangement

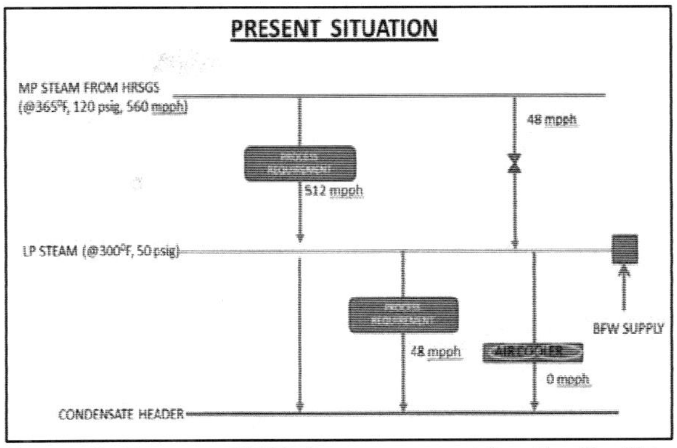

Figure 4: Present CHP Arrangement for the facility

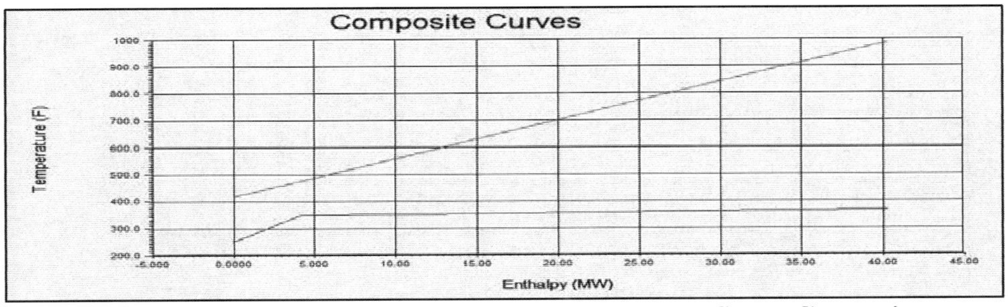

Figure 5: Composite curves of Exhaust Gases for MP Steam Generation

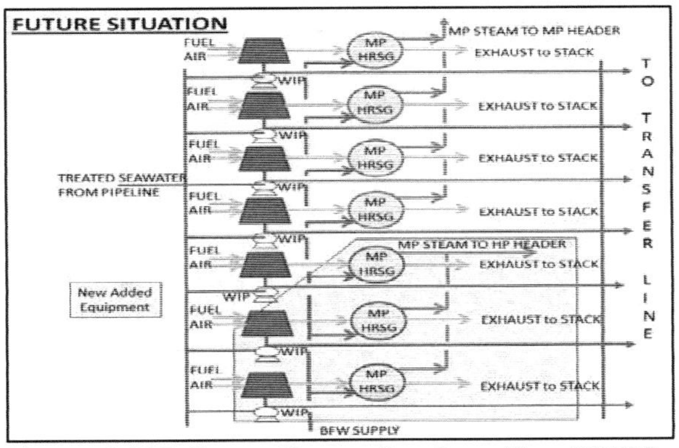

Figure 6: MP Steam Generation from New HRSGs

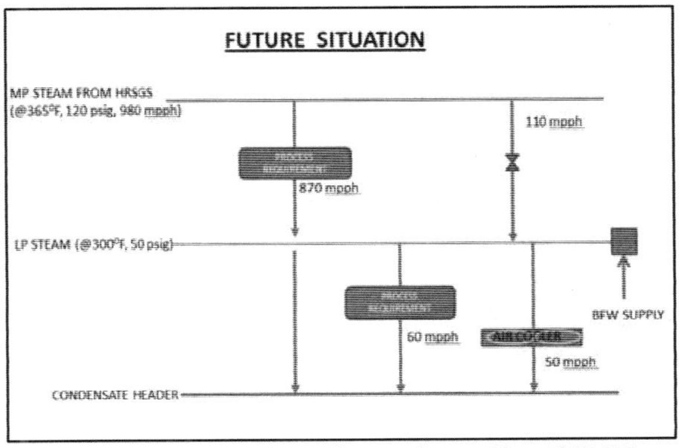

Figure 7: CHP for MP Steam from New HRSGs

If same arrangement is followed (i.e. same amount of MP steam is produced in the new HRSGs) in the future expansion which is considered as base case to compare all other alternatives schemes, not a perfect steam generation-consumption match is possible (figure 7). There is a need to optimize the system, as explained earlier the method used is to target the exhaust gases to produce steam and facility CHP model is utilize to evaluate all possible schemes. Targeting of exhaust gas (as shown in figure 5 & 8 Composite Curves for the system) shows that there is potential to produce HP steam of same amount (140 mpph per HRSGs) instead of MP steam which can be let down to produce power. With this information there are two cases which can be proposed as options to design the facility optimal steam system.

1. New HRSGs can produce HP steam instead of MP steam and
2. New HRSGs as well as existing HRSGs can be upgraded to produce HP steam

In the first option the new design HP HRSGs cost need to justify that additional investment for HP versus MP HRSGs while in second option cost of new HP HRSGs as well as cost of upgrading existing HRSGs will need to justify additional power production. These are the options which need to be optimized through evaluation of Capital Expenditure (Capex) and Operating Expenditure (Opex) to get results in terms of Net Present Value (NPV).

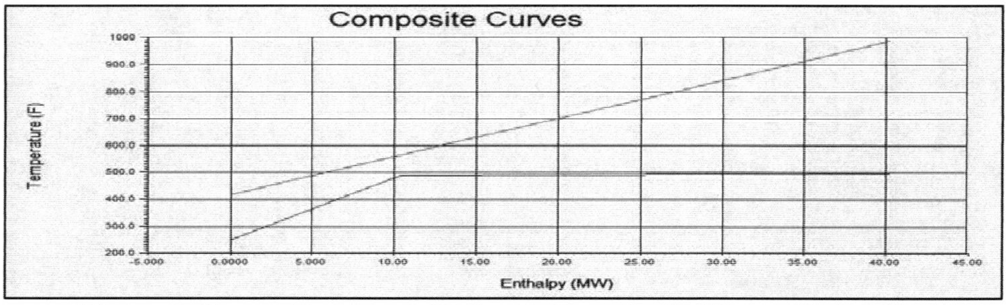

Figure 8: Composite curves of Exhaust Gases for HP Steam Generation

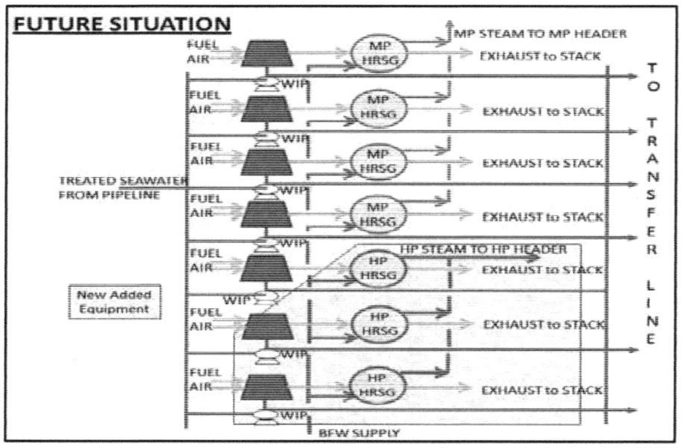

Figure 9: HP Steam Generation from New HRSGs

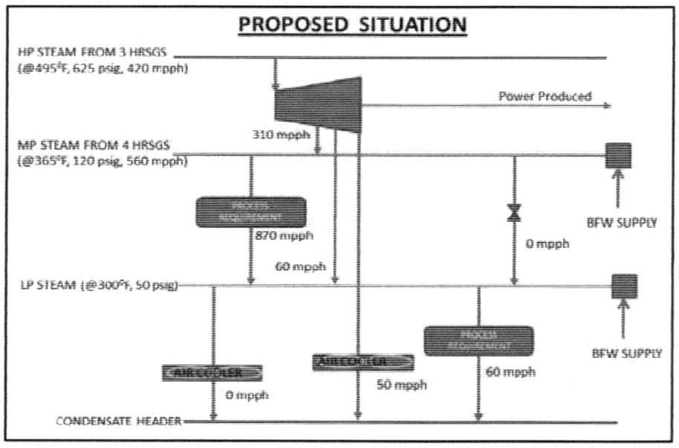

Figure 10: CHP for HP Steam from New HRSGs

Both options are quite similar with only difference, that in the first option we produce HP steam only with new HRSGs and have limited power production. Second option, with the upgrade of existing HRSGs we can produce more power but needs extra capital to upgrade the existing HRSGs to produce HP steam.

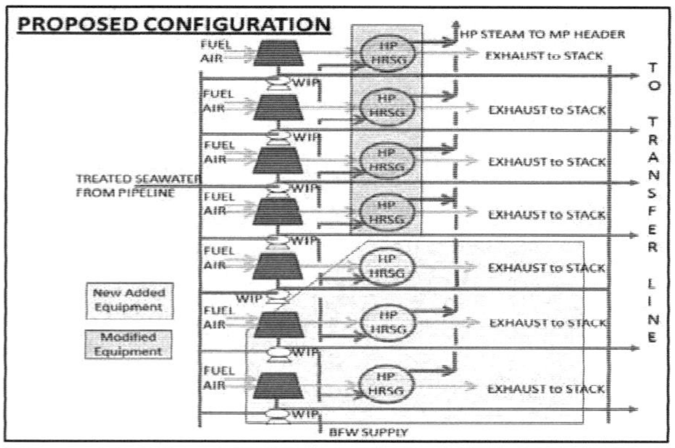

Figure 11: Max. HP Steam Generation Arrangement

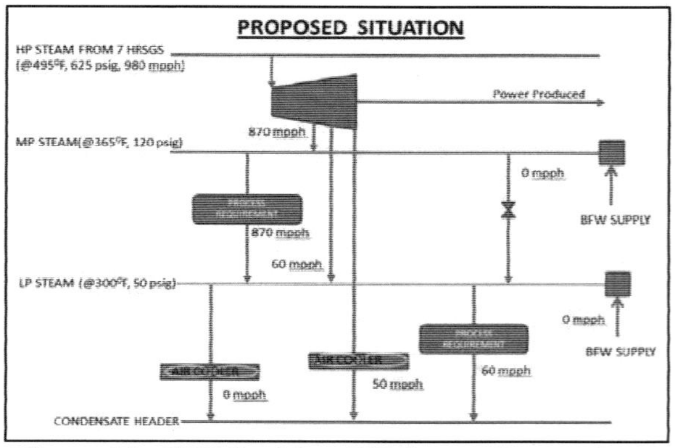

Figure 12: CHP for Max. HP Steam Generation

Of course there might be some options which can be proposed based on the utilization of the LP steam for some of the application requiring chilling but the quantity of excess steam is small to completely satisfy duty require by any chillers. Taking part load from the mechanical chillers will results in two systems which is very expensive especially for the same temperature range. Thus we can conclude that only these two options available to be evaluated in terms of NPV for the optimization.

Results and Discussion

As the exhaust temperature of the flue gas is very high (418°F) and needs to re-consider as we are using sweet gas as fuel having dew point below (250°F). The constraint of 400°F is due to utilization of sour gas in case of emergency when sales gas is not available, rarely occurring case (no incident till date from the plant start-up, more than 5 years). This lead to design a system utilizing heat wasted in flue gas and it was proposed to have a system which utilize the heat and switch back to exhaust flue gas at higher temperature when sour gas is in service. The proposed design in figure 14 can fulfil the requirement, though the proposed design is shown with blower but it could be designed without blower and depend on actual site arrangement and HRSGs in service etc. In this design flue gas leaving HRSGs shall be reduced to below 300°F if the fuel used is sweet gas but it would by-pass one of the HRSGs to be exhausted at above 400°F if the fuel gas used is sour gas. Targeting with new condition shows that we can produce additional 30 mpph MP steam per HRSGs (i.e. 170 mpph instead of 140 mpph) or we can produce 140 mpph per of HP steam and 30 mpph of MP steam from exhaust of each WIP.

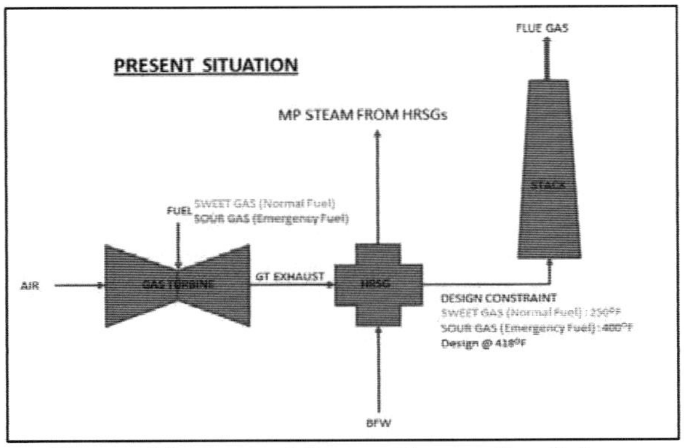

Figure 13: Present Steam Generation Scheme

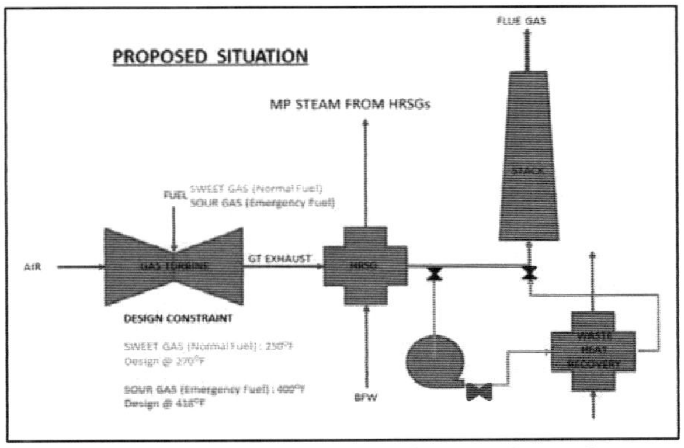

Figure 14: Proposed Steam Generation scheme

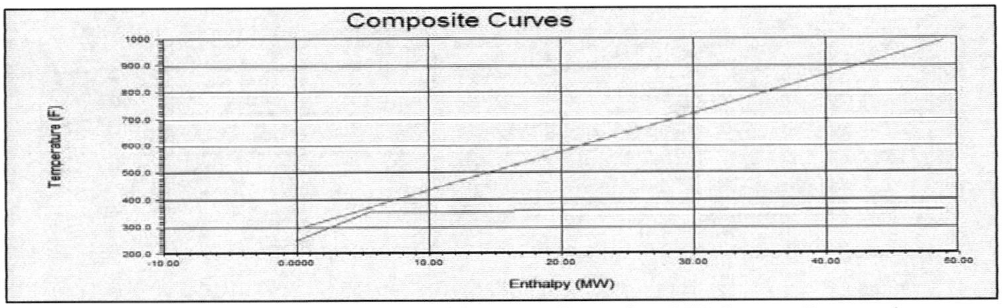

Figure 15: Composite curves of Exhaust Gases for MP Steam Generation

Based on the targeting results, there are possibly many cases which need to be evaluated for selecting optimum case, based on steam production following may be some additional options (in addition to earlier cases) for selecting optimum.

1. No change in existing HRSGs and new HRSGs produce more MP (170 mpph per HRSGs) steam
2. No change in existing HRSGs and new HRSGs can produce HP steam (140 mpph) and MP steam (30 mpph)
3. Modified existing HRSGs as well as new HRSGs to produce MP (170 mpph per HRSGs) steam
4. Modified existing HRSGs as well as new HRSGs to produce HP steam (140 mpph) and MP steam (30 mpph).

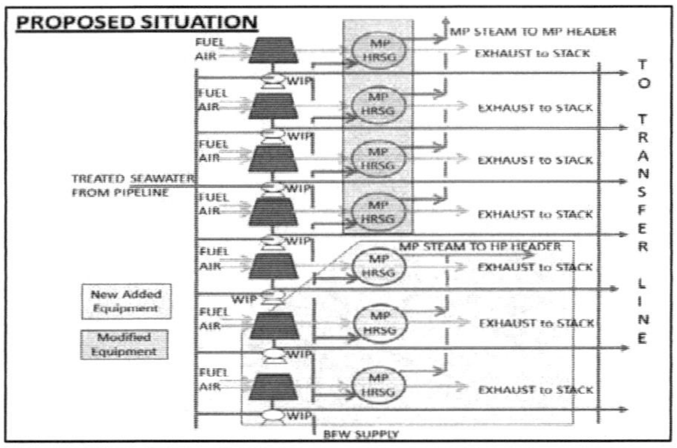

Figure 16: Max. MP Steam Generation Arrangement

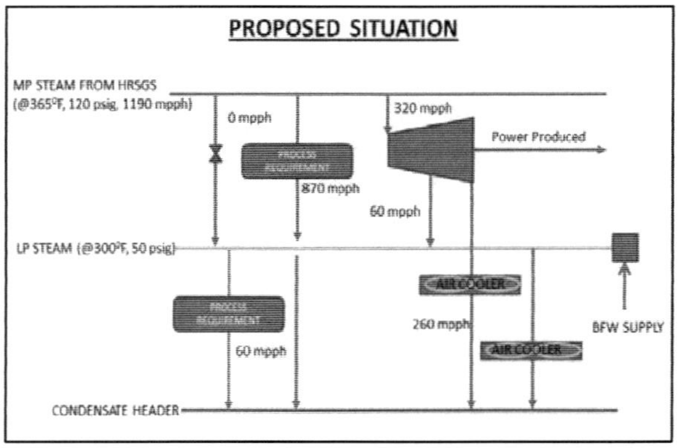

Figure 17: CHP for Max. MP Steam Generation

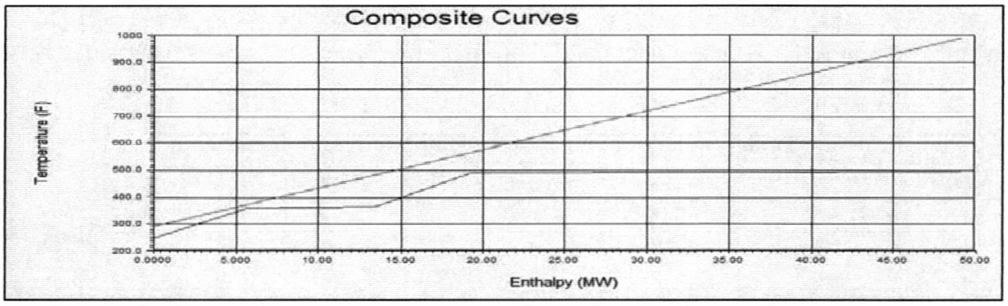

Figure 18: Composite curves of Exhaust Gases for HP & MP Steam Generation

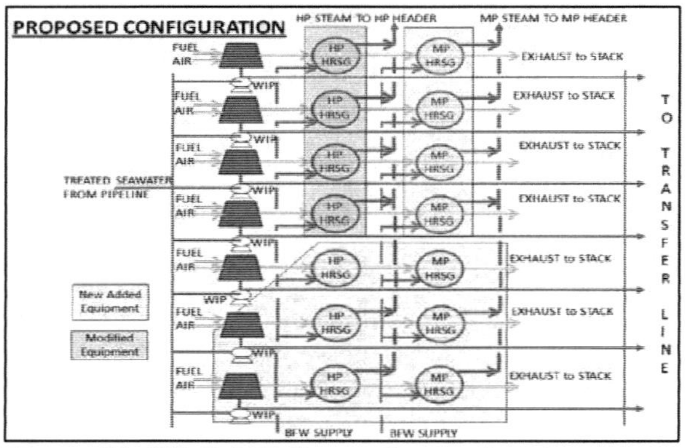

Figure 19: HP & MP Steam Generation Arrangement

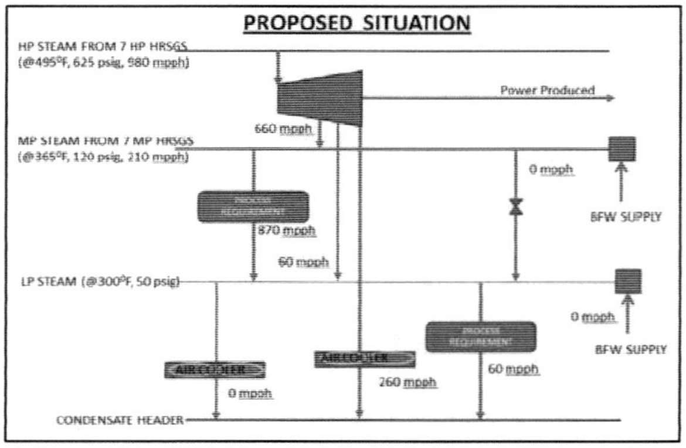

Figure 20: CHP for HP & MP Steam Generation

Now considering these cases clearly shows that power production from various options can vary from 0 (base case) to 14 MW (case shown in figure 19), which means different opex for each case. It also shows that the lowest investment cost is for the base case (i.e. putting same HRSGs design as the existing HRSGs) and highest capex for case shown in figure 19. These means all cases have potentials to be winning case depending on the price of the power produced versus capital required to produce it.

All the presented cases are considered with the assumption that the available heat shall be utilized to produce power (site power demand) and also with the assumptions that steam turbine is the best technology to do it. It is instructive to mention that even now optimization involving boilers to produce steam to use it as driver for one WIP can be explored. But utility system should complement process system and vice-versa, demand analysis results shows that along power requirement there is lot of refrigeration requirement for the site.

Site Wide Impact

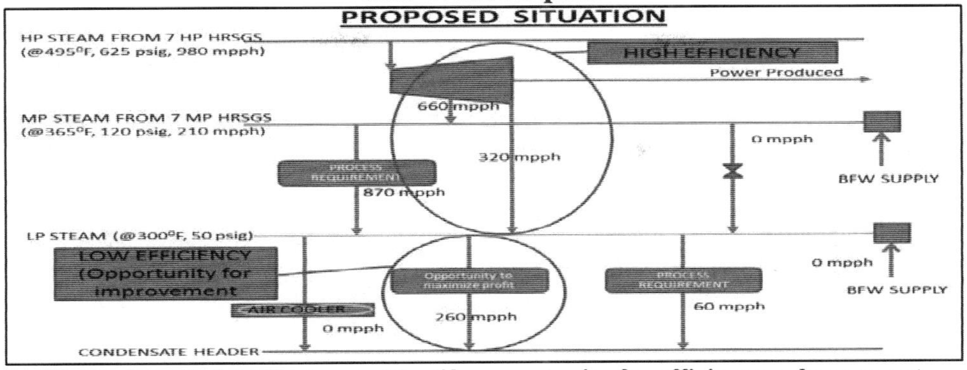

Figure 21: CHP utilization to identify opportunity for efficiency enhancement

Now considering site-wide impact for the proposed design, it is evident from the figure 21 that high grade heat utilization efficiency is quite high (back pressure turbine) can be further increase marginally. Like utilizing the available heat to produce steam and instead of utilizing steam to produce power, use it as steam driver for rotating equipments in the plant. It would be very interesting problem if the steam produced is enough to drive at least one WIP as this is not the case and there is not much potential to improve in utilization of high grade energy. Still there are lots of opportunities in the utilization of low grade energy to produce power via available technology like Organic Rankine Cycle (ORC), Kalina Cycle etc. As mention in Table 1 the amount of power produced from different technology is not very different for different technologies, which lead to the conclusion that initial investment or capex is the deciding factor for their utilization. Analysis of utility demand from the processes clearly indicates that there is lot of refrigeration requirement from the process which could be fulfilled by the available low grade energy utilizing Absorption Chillers or other available technologies to maximize the savings for the site.

Technology	Power Output
Condensing Turbine	~11 MW
Organic Rankine Cycle (ORC)	~7 MW
Kalina Cycle	~7-8 MW

Table 1: Utilization of Low Grade Heat for Power Production

Absorption chillers are systems which utilize heat to produce chilling capacity for low temperature cooling. The available steam quantity with new design constraints is more than enough for satisfying all chilling load for gas trains and inlet cooling for GTs. The amount of steam left after fulfilling facility chilling requirement is used to produce power via condensing turbine. Calculations are provided to show that the amount of savings increases many fold from initial saving just by power production.

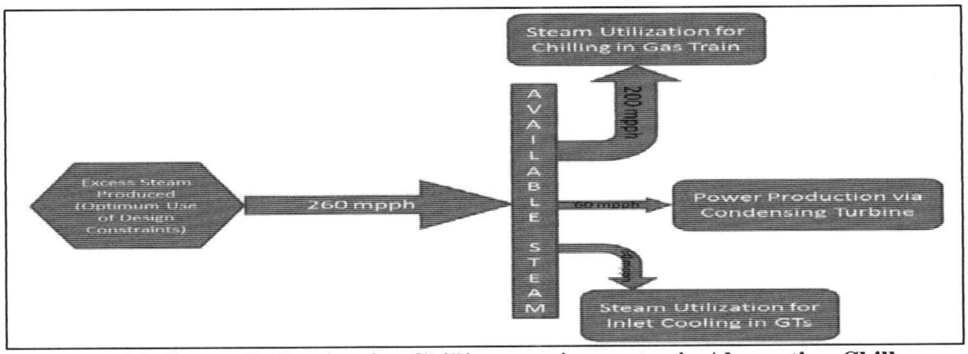

Figure 22: Steam Utilization for Chilling requirements via Absorption Chillers

Steam Utilization for Gas Train

Requirement by Trains & Stages	Operating Cooling		Design Cooling	
	MMBtu/hr	RT	MMBtu/hr	RT
First Stage Cooling for 1st Gas Train	55.73	4644	65.6	5467
First Stage Cooling for 2nd Gas Train	55.73	4644	65.6	5467
First Stage Cooling for 3rd Gas Train	27.87	2322	52.8	4400
Total First Stage Cooling	139.33	11610	184.0	15333
Second Stage Cooling for 1st Gas Train	23.53	1961	25.9	2158
Second Stage Cooling for 2nd Gas Train	23.53	1961	25.9	2158
Second Stage Cooling for 3rd Gas Train	11.77	980	21.0	1750
Total Second Stage Cooling	58.83	4902	72.80	6067
Total Cooling	**198**	**16513**	**257**	**21400**

Table 2: Cooling requirement for Gas Trains

Steam Utilization using absorption chiller	Steam Consumed (lb/hr)	Refrigeration (RT)
Refrigeration for 1st Stage for 2 Nos Existing Gas Train	77753	9288
Refrigeration for 2nd Stage for 2 Nos Existing Gas Train	32828	3922
Refrigeration for 1st Stage for New Gas Train	19438	2322
Refrigeration for 2nd Stage for New Gas Train	8207	980
Total Steam Utilized	**138226**	**16513**
Water Consumption	**m^3/hr**	**gpm**
Cooling Water Required (Refrigeration)	12192	53681
	m^3/hr	**gpm**
Make-up Water Required	**183**	**805**
Electrical Power Saving	**HP**	**kW**
E-51 (2fans @50 hp each) (New)	100	75
E-54A (6 fans with 40 hp each) (New)	240	179
E-54A (6 fans with 50 hp each) (New)	300	224
Refrigeration Compressor power for 2 Nos Existing Gas Train	25480	19000
Refrigeration Condenser power for 2 Nos Existing Gas Train	3840	2863
Refrigeration Compressor power for New Gas Train	9650	7196
Refrigeration Condenser power for New Gas Train	1440	1074
Total Power Saved	**41050**	**30611**
Power Consumption	**HP**	**kW**
Hot Water circulation pump (1pump @ 175hp) (New)	175	130
Cooling Water Tower Circ.Pumps (2 pump @ 750 hp) (New)	1500	1119
Cooling Water Circulation Pumps (2 pump @ 1500 hp) (New)	3000	2237
Cooling Water fans (20 fans @ 30 hp) (New)	600	447
Total Power consumption	**5275**	**3934**
	HP	**kW**
Net Power Saving in site Import	**35775**	**26677**
	HPH	**kWH**
Net Power Saving in site Import/year (@8760 operating hours)	**313389000**	**233694146**

Table 3: Calculation of potential Operating Cost Saving utilizing Absorption Chiller for Gas Trains

In table 2 gas train chilling requirement is shown and in table 3 the amount of power savings shown for case utilizing Absorption Chillers as compared to Vapor Compression system and the power saved is about 27 MW and mainly due to high power consumption in Propane Refrigeration System.

Steam Utilization for Inlet Cooling

Air Flow at ISO condition for 1 GT (MS5002D)	1113000	lb/hr
Energy Required to cool air for 1 GT from 104°F to 59°F	12.14	MMBTu/hr
Air Flow at ISO condition for 7 GT (MS5002D)	7791000	lb/hr
Energy Required to cool air for 7 GT from 104°F to 59°F	84.98	MMBTu/hr
Total Energy Required to cool all air from 104°F to 59°F	84.98	MMBTu/hr
Total Energy Required to cool all air from 104°F to 59°F	7082	RT
Steam Required for the cooling	59281	lb/hr
Water Consumption	**m^3/hr**	**gpm**
Cooling Water Required (Refrigeration)	5229	23022
Total Cooling Water Circulation Required	**5229**	**23022**
Make-up Water Required	**m^3/hr**	**gpm**
Make-up water for Cooling Tower	261	1151

Power Produced	HP	kW
Extra Power Generation (Inlet Air Cooling for 6 GTs and 1 Cogen)	72482	54050
Power Consumed	**HP**	**kW**
BFW pump (4pump @ 175hp each) (New)	700	522
Cooling Water Tower Circ.Pumps (2 pump @ 900 hp) (New)	182	136
Cooling Water Circulation Pumps (2 pump @ 1200 hp)(New)	2400	1790
Cooling Water fans (20 fans @ 25 hp) (New)	500	373
Flue Gas Blower (3 for three HRSGs @ 750 hp each)	2250	1678
Total Power Consumed	6032	4498
Fuel Consumption Increase (lb/sec)	5.48	
LHV for Fuel Gas (Btu/lb)	18310	
Fuel Consumption Increase (lb/year) (@8760 operating hours)	127837440	
Increase Energy Consumption (MMBtu/yr @8760 operating hours)	**2340704**	
Net Savings	**HP**	**kW**
Net Power Saving in site Import	**66450**	**49552**
	HPH	**kWH**
Net Power Saving in site Import/year (@8760 operating hours)	**430597563**	**321096560**

Table 4: Calculation of potential Operating Cost Saving utilizing Absorption Chiller for Inlet Cooling

Table 4 shows the effect of GTs inlet cooling which is saving of about 50 MW but with increase of 5.5 lb/sec increase in Fuel Gas. There are other technologies that can be utilize to produce similar results and might be options for optimization problem. This case study demonstrated that how changing a design constraint can lead to several options for optimization and consequently improving the efficiency of the overall system.

Results and Conclusions

In many industrial systems the design optimization is necessary. A lot of margins in adapted designs can lead to low overall system efficiency. Several beneficial process schemes may be unexplored if proper design limitations are not utilized by the designer. The system design limitation identification is very important to specify the proper process/utility system synthesis. In the WIPs case study presented in this paper, it is demonstrated that the system overall efficiency can be improved by designing system to exploit all the available energy in the flue gas. The extracted waste energy with the available Absorption Technology provided an opportunity to enhance the utilization of the available energy customized to the site demand. This enhanced overall system efficiency can only be attained via the proper consideration of the system limitations and designing system to those limitations.

References

Linnhoff, B., and Hindmarsh, E., "The pinch design method for heat exchanger networks" Chemical Engineering Science, Vol. 38, PP 745-763, (1983).

Linnhoff, B., "Pinch analysis-state of the art overview" Transactions Institution Chemical Engineering Research and Design, Vol. 71(A5), PP 503-522, (1993)

Biegler, L.T., Grossmann, I.E., Westerberg, A.W., "Systematic Methods of chemical process Design" Prentice Hall, New Jersey (1997)

Noureldin, M.B., "Methods for Heat Exchanger Network Energy Efficiency Assessment and Lifetime Retrofit", USP 8,150,560, (2012)

Noureldin, M.B., "Systems and Program Product for Heat Exchanger Network Energy Efficiency Assessment and Lifetime Retrofit", USP 8,150,559, (2012)

N.D.K. Asanta, X.X. Zhu, An automated approach for heat exchanger network retrofits featuring minimal topology modification, Computer & Chemical Engineering 20 (1996) 7-12

B.L. Yeap, D.I. Wilson, G.T. Polley, S.J. Pugh, Mitigation of crude oil refinery heat exchanger fouling through retrofits based on thermohydraulic fouling models, ChED 82A (2004) 53-71

399612 Control of Processes Whose Optimal Operating Conditions Are on the Saturation Edge

Abstract:

Hydrocarbon reforming can be used to produce economically hydrogen, a source for the proton exchange membrane fuel cell. The hydrogen-rich gas from hydrocarbon reforming reactors can have CO up to several thousand ppm level which poisons the anode catalyst of the fuel cell. The CO concentration need be removed below 10ppm. The preferential oxidation (PROX) is one of the simplest ways for this purpose. It uses the following oxidations:

$CO+0.5O_2$ --> CO_2

$H_2+0.5O_2$ --> H_2O

In above reactions, the CO oxidation should occur preferentially over the H_2 oxidation. The copper-cerium oxide catalyst of $CuO-CeO_2$ has been reported to have high activity and selectivity for CO oxidation. The CO PROX reactors with $CuO-CeO_2$ catalyst can remove CO from several thousand ppm to less than 10ppm. To design this PROX reactor, kinetics have been obtained by Lee and Kim ("Kinetics of CO and H_2 oxidation over $CuO-CeO_2$ catalyst in H_2 mixtures with CO_2 and H_2O," *Catal Today* 2008;132;109-116). Rigorous simulations have been done by Kim et al. ("Preferential CO oxidation over $CuO-CeO_2$ in excess hydrogen: Effective factors of catalyst particles and temperature window for CO removal," *Int. J. Hydrogen Energy* 2013;38;4429-4436). In this PROX reactor, limited O_2 should be added and O_2 should not exist in the exit stream for the optimal operations. A large amount of O_2 feed results in O_2 in the exit stream, reducing the efficiency of the PROX reactor. A small amount of O_2 feed results in high CO concentration in the exit stream and the objectives of the PROX reactor will not be met. The oxygen feed should be controlled on the edge just before O_2 appears in the exit stream.

There are several processes where optimal operational conditions are on the saturation edge as shown in the above PROX reactor. For example, the reactor temperature of polymerization reactor is given by the vaporization temperature of monomer and the optimal heat input is such that the reactor temperature should be on this saturation temperature edge. Additional heat increases the boil-up amount of monomer and the additional amount of cooling water should be needed. Some recycle systems with excess and limiting reagents give these control problems that maintain operating points on the saturation edges.

When the controlled variable is saturated, the variable has no information for feedback and the conventional control cannot be used. Control methods based on the sliding mode method and extremum seeking control method with coordinate system transformations are suggested and their performances are compared.

SYNTHESIS AND OPTIMIZATION OF AN INTEGRATED WATER AND MEMBRANE NETWORK SYSTEM WITH MULTIPLE ELECTRODIALYSIS REGENERATORS

Nielsen Y. Mafukidze, University of the Witwatersrand, Johannesburg, South Africa

Thokozani Majozi, University of the Witwatersrand, Johannesburg, South Africa*

Abstract

The shrinking supplies of freshwater globally, coupled with strict environmental regulations, have driven the manufacturing industry towards sustainable water management for the minimization of freshwater intake and wastewater generation. By using process integration and its enabling tools, this work considers the synthesis of an optimal water network with multiple regeneration capabilities. Development of the proposed framework is achieved by embedding a subnetwork of detailed electrodialysis models within a water network. Based on a superstructure and fixed-flowrate, the optimization problem is formulated as an MINLP model and solved in GAMS/DICOPT. To demonstrate the applicability of the proposed mathematical model a literature case study on a pulp and paper plant is presented and the results indicate a potential 12% savings in freshwater intake, 16% reduction in wastewater generated and a 14% saving in the total annualised cost for the entire network.

Keywords: Sustainable; Synthesis; Optimization; Electrodialysis

1. Introduction

Sustainability of human activities, primarily production and consumption, has been of growing concern in the process industry. Among the sustainability mega-forces, water scarcity has risen to the top of the corporate agenda for most entities in the industrial sector. In the face of this challenge, a number of industries are taking measures to become stewards of this vital resource through sustainable water management. With regards to the minimisation of freshwater intake and wastewater generation, process integration has presented enabling tools for sustainable water network design such as water recycle, reuse as well as regeneration-recycle and regeneration-reuse [1]. Optimization of water networks has become an essential instrument of sustainability as it allows for the realization of the full benefits, both economic and environmental, of using the aforementioned enabling tools.

The two optimisation approaches that remain irrefutable are the insight-based 'pinch' approach and mathematical programming. Most of the early work reported on water network optimisation was insight-based, which being graphically-based, focused on flowrate targeting and network design methods. Over the years, the technique has been refined and modified to extend its applicability to various systems found in the real world. Noteworthy is that these graphical methods offer low computation expense but are however limited to mass transfer based operations and grassroots design purposes [2]. Mathematical programming techniques involve the development of models based on mathematical relations that describe the system

* Corresponding author. Tel.: (+27) 11 717 7567/7384
Email address: Thokozani.Majozi@wits.ac.za

under analysis and the application of rigorous algorithms to obtain solutions. This technique became more prominent much later than pinch, as researchers realised its capability of treating rigorous, large-scale, complex systems, which pinch analysis is incapable of handling [2].

Within water networks, regeneration allows for the partial treatment of wastewater before being recycled or reused. An interesting observation is that for most of the work done so far on water-regeneration network optimization, the technologies of choice, number of regeneration units, arrangement of the regeneration train and regeneration design parameters are not set as decision variables; rather they are assumed to be known or oversimplified in a 'blackbox' fashion [3]. Contrary to that, Rangaiah and Wei [3] highlight that it is imperative to consider the technicalities of the system, including regeneration, such that complete structural and parametric (design) optimization of the given system is satisfied. Only a few researchers have considered the use detailed regenerator models for water networks. This includes El-Halwagi [4], Galan and Grossmann [5]. Khor *et al.*, [6] and more recently Yang *et al.* [7].

Electrodialysis (ED), the regeneration technology adapted in this work, has found large industrial scale use in the production of high quality industrial water as well as wastewater treatment of industrial effluents [8]. This water treatment process is based on the electro-migration and selective transport of ionic species induced by an electric current through a series of alternately arranged anion and cation-exchange membranes [8]. A majority of the insight on this particular process has been presented by Strathmann [8]. The current paper addresses the synthesis of a comprehensive water and membrane network that simultaneously gives an optimal design for the ED subnetwork. This is achieved by embedding a subnetwork of detailed ED models into a water network problem. Each regenerator model incorporates detailed cost correlations that are functions of the design variables.

2. Problem statement

In this work we consider a single-contaminant, integrated water network with multiple electrodialysis regenerators. The problem can be stated as follows. Given:

1. A set of fixed-flowrate water sources, J, with known flowrates and concentration of contaminant, which are flexible to recycle or reuse.
2. A freshwater source, FW, with variable flowrate and known contaminant concentration.
3. A set of fixed-flowrate water sinks, I, with known flowrates requirements and maximum allowable concentration of contaminants.
4. A wastewater sink, WW, with variable flowrate and maximum allowable concentration of contaminant. Effluent that requires further contaminant removal by chemical, biological or other means goes to this sink.
5. A set of regenerators, R, of different performances in terms of the removal ratio and the maximum allowable concentrations in to the regenerators. Flowrate into the regenerators is variable. Also given are some design parameters for ED units.

The objective is to synthesise an optimal integrated water network and to determine the design of the ED subnetwork that corresponds with the optimal water network. The optimal configuration is one that minimises freshwater use, generation of wastewater, regeneration energy and the capital and operating costs of both regeneration and the piping interconnections.

3. Optimization model

3.1. Superstructure representation

A superstructure approach that considers the water sources sinks and regeneration subnetwork simultaneously was used to formulate the water network problem considered in this work. The proposed superstructure includes all feasible interconnections between different units in the water-regeneration network and is presented in Figure 1. It includes a regeneration subnetwork as part of the water network. Provisions have been made for both parallel and series connection of regenerators as well as allowance for recycle of regenerator products back to the same regenerator feed. These provisions allow for a complete synthesis within the regeneration subnetwork and water network.

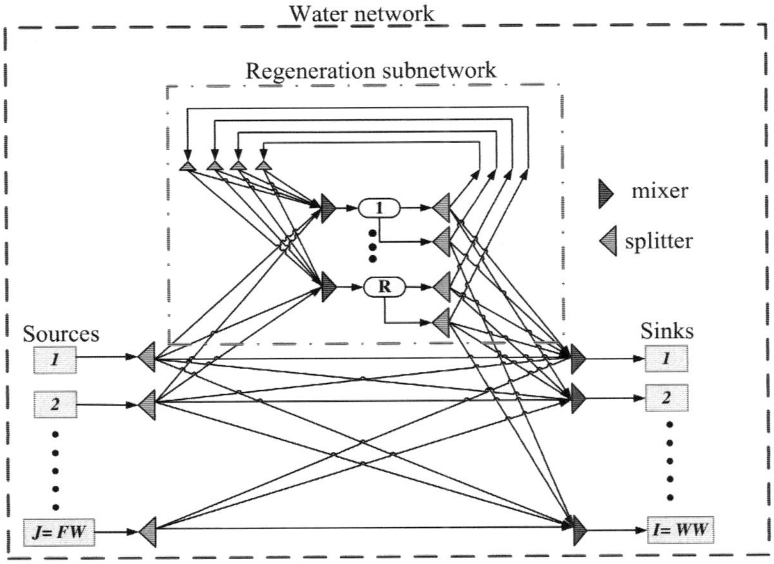

Figure 1: Superstructure representation of the water network.

3.2. Mechanistic electrodialysis regeneration network model

This section presents a mechanistic model formulation for a detailed design of an electrodialysis regeneration unit based on Tsiakis and Papageorgiou, [9]. The formulation includes mathematical equations for current required, voltage, membrane area, number of cell pairs, stack length, pumping energy and desalination energy. The regeneration network involves both capital and operational costs. The representative total annualised cost (*TAC*) function for the regeneration network is given by Equation (1). This function is included in the overall objective function of the water network, such that the energy consumption and subsequent cost of regeneration are minimised in conjunction with water consumption. To further improve flexibility, the model is enabled to only select the necessary regenerators that result in an optimal solution. This is achieved by prefixing the TAC function with a binary variable, y_r^{ED} which becomes zero when the respective unit is not activated.

$$TAC_r = y_r^{ED}\left[A_r\left(\frac{k^{mb}}{n}\right) + t^d k^{el} Q_r^{dil}\left(E_r^{spec} + E_r^{pump}\right)\right] \qquad \forall r \in R \qquad \text{[Eq. 1]}$$

3.3. Water Balances

To ensure connectivity between the regeneration network, the sources and the sinks, water balances are established based on the superstructure presented in Figure 1. These show how water from the different units split/mix into the other units. The maximum allowable load that each sink and regenerator tolerates is taken into account by means of contaminant balances. Logical constraints that ensure that streams from the same regenerator are not re-mixed in a single sink or recycled into the same regenerator are introduced. Additionally, to ensure that the maximum flowrates that can be accommodated by the piping are not exceeded, appropriate bounds are allocated to all the flowrates. These bounds are chosen based on practically feasible flowrates, such that any stream with a flowrate beyond these limits is not allowed to exist.

3.4. Objective function

The objective is to minimise the total annualised cost of the water network, which comprises the freshwater cost, wastewater treatment cost, annualised regeneration cost, as well as the capital and operating cost of piping. To account for capital and operational cost of piping, the length of pipe run, material of construction and linear velocities within the pipes were considered. The formulation of the objective function is expressed in Equation (2), where A^f is an annualisation factor.

$$obj = \min \left\{ \begin{array}{l} \left(C^{FW} FW + C^{WW} WW\right) t^d + \sum_{r=1}^{R} TAC_r + \\ A^f \left[\begin{array}{l} \sum_{j=1}^{J}\sum_{i=1}^{I} D_{j,i}^s \left(\frac{pQ_{j,i}^s}{3600 \cdot v} + q \cdot y_{j,i}^s \right) + \sum_{r=1}^{R}\sum_{i=1}^{I} D_{r,i}^d \left(\frac{pQ_{r,i}^c}{3600 \cdot v} + q \cdot y_{r,i}^c \right) + \\ \sum_{r=1}^{R}\sum_{i=1}^{I} D_{r,i}^d \left(\frac{pQ_{r,i}^d}{3600 \cdot v} + q \cdot y_{r,i}^d \right) + \sum_{r=1}^{R}\sum_{r'=1}^{R} D_{r,r'}^y \left(\frac{pQ_{r,r'}^x}{3600 \cdot v} + q \cdot y_{r,r'}^x \right) + \\ \sum_{r=1}^{R}\sum_{r'=1}^{R} D_{r,r'}^y \left(\frac{pQ_{r,r'}^y}{3600 \cdot v} + q \cdot y_{r,r'}^y \right) + \sum_{j=1}^{J}\sum_{r=1}^{R} D_{j,r}^k \left(\frac{pQ_{j,r}^k}{3600 \cdot v} + q \cdot y_{j,r}^k \right) \end{array} \right] \end{array} \right\} \qquad \text{[Eq. 2]}$$

It is assumed that all the pipes share the same p and q parameter properties, stream velocity, v, and 1-norm Manhattan distance. The resulting mathematical model is a mixed integer nonlinear programming (MINLP) problem that was solved using GAMS/DICOPT.

4. Illustrative Example

The mathematical model developed is applied to a literature based pulp and paper plant case study [10]. This case study was chosen mainly because the industry is known to produce a lot of ionic effluent. This wastewater comes mainly from the bleaching section of the plant, where water is contaminated with chlorides and chlorates.

Table 1: Limiting data for the water sources and sinks.

Source, *j*			Sinks, *i*		
j	**Flowrate (*t/h*)**	**Concentration (*mg/L*)**	*i*	**Flowrate (*t/h*)**	**Max. concentration (*mg/L*)**
1	890.1	0.003	1	1399.5	34.0
2	145.0	269.6	2	145.0	0.00
3	102.4	0.014	3	576.2	0.37
4	3099.0	499.8	4	3099.0	0.00
FW	∞	0	*WW*	∞	398.3

Table 1 shows the plant limiting data in terms of flowrate and contaminant (Cl) concentration for the sources and sinks amenable for wastewater integration. According to this table, the plant water network involves five water sources including the freshwater source and five sinks including the wastewater discharge.

To ensure a detailed analysis of the model, four scenarios are considered. Scenario 1 is water integration without regeneration. Scenario 2 is the case where water integration with regeneration involving two detailed ED regenerators is considered. However in this case, series connection and recycle within the regeneration subnetwork are forbidden just as in the case presented by Khor, *et al.*, [6]. Scenario 3 considers a water network with one detailed ED regenerator with capabilities of recycle within the regeneration subnetwork. Note that within this scenario two cases were considered, that is the case where the removal ratio of the regenerator is fixed at 0.733 and the case where it is variable. For the case where the removal ratio is fixed, an arbitrary value is chosen. Scenario 4 is similar to scenario 3 except that it has the capability of using two regenerators. Fixed removal ratios at 0.733 and 0.950 for the first and second regenerators respectively are chosen for the fixed removal ratio case.

Comparisons of the presented cases show that scenario 1 is the worst case possible. Scenario 4 proved to be the best and most preferable, since the largest amount of water was treated at the lowest total cost thereby minimizing the freshwater intake and wastewater generation.

Table **2** presents the optimum results for scenarios 4. It is evident from the results that by allowing the solver to choose the optimal configuration within the regeneration subnetwork in terms of recycles, series and parallel connections as well as making the removal ratio and number of regenerators required variable, there exist possibilities of obtaining better solutions rather than fixing them arbitrarily beforehand. In this way the designer gains more control of the unit performance by being able to stipulate the required membrane characteristics to the manufacturer. Additionally, since the technicalities of regeneration are included in the model, the performance of the regenerator, dictated by the removal ratio, directly influences the quantity of wastewater to be treated. As such allowing the optimiser to select both the performance and the quantity of water to be treated means a better optimum is guaranteed.

Table 2: Optimum results for water network for scenario 4.

	Fixed *RR*	Variable *RR*
Freshwater flowrate (*t/s*)	1.006	0.989
Wastewater generation (*t/s*)	0.733	0.716
Total water regenerated (*t/s*)	0.699	0.586
Regeneration cost (million $)	0.562	0.353
Total cost (million $)	50.2	48.9
Optimal RR^*	-	r_1= 0.779 r_2= 0.770
CPU time (*s*)	6129	17230

The results also show that incorporating multiple regenerators can increase the chances of getting a better optimal solution, as long as the regenerators are not forced into the system. It is noteworthy that the proposed model is not limited to only two ED regenerators but can accommodate more as deemed necessary by the designer.

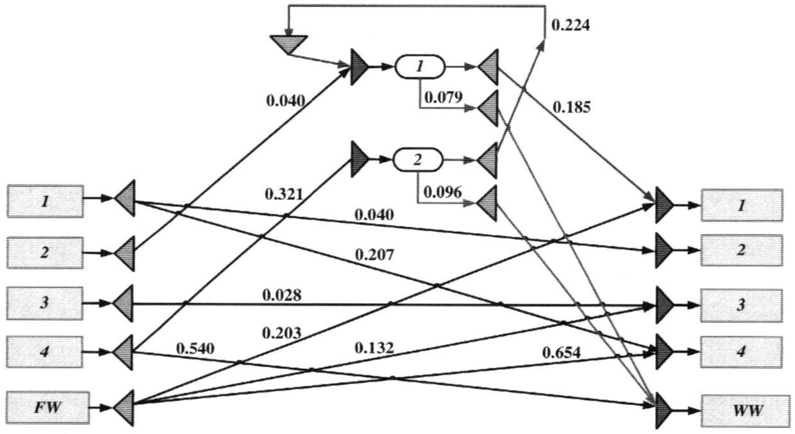

Figure 2: Optimum water network configuration for scenario 4 (variable RR).

Figure 2 shows the optimal network configuration and flowrates obtained for scenario 4. All flowrates are reported in t/s. According to Figure 2, that both series and parallel connection are vital for optimality. With this configuration in place, the plant is capable of generating savings of up to 12.7% in freshwater intake, reduction of up to 16.2% in wastewater generation and a 14.1% saving in the total annualised cost compared to the worst case scenario. The optimal regenerator design suggests an installation of two regenerators, r_1 and r_2, one with 2880 cell pairs and the other with 4356, amounting to a total membrane area of 3837 m^2 and 3921 m^2 respectively.

5. Conclusion

This work addresses the synthesis of a multi-regeneration water network by proposing an MINLP model that incorporates a detailed model of an electrodialysis regeneration subnetwork within a water network. To demonstrate the applicability of the proposed mathematical model, it is applied to a literature case study and solved using

GAMS/DICOPT. The results show that the use of a detailed model guarantees more accurate and reliable results in terms of water network synthesis as well as regenerator design. However, the nature and structure of the model proposed is quite complex, thereby resulting in very large computational times being incurred. Additionally, the system considered in this manuscript was limited to a single, ionic contaminant, whereas industrial effluent often contains multiple ionic contaminants as well as multiple non-ionic contaminants. As such, offering scope for future work, it would be desirable to extend the problem to allow for multiple contaminants, both ionic and non-ionic, using multiple regeneration technologies in order to be able to select the best available technology for the different streams of wastewater.

6. Nomenclature

Sets

J = $\{j|j = \text{water sources}\}$

I = $\{i|i = \text{water sinks}\}$

R = $\{r|r = \text{regenerators}\}$

Parameters

k^{el}	cost of electricity
k^{mb}	cost of membrane
C^{FW}	freshwater cost
C^{WW}	wastewater treatment cost
A'	annualisation factor
n	maximum equipment life
t^{d}	operating time per year
p	parameter for carbon steel piping
q	parameter for carbon steel piping
v	pipe linear velocity
$D^{s}_{j,i}$	Manhattan distance between source, j and sink i
$D^{d}_{r,i}$	Manhattan distance between regenerator, r and sink, i
$D^{y}_{r,r'}$	Manhattan distance between regenerators r and r'
$D^{k}_{j,r}$	Manhattan distance between source, j and regenerator, r

Continuous variables

TAC_r	total annualized cost for regenerator, r.
A_r	membrane area required by regenerator r.
E^{spec}_r	specific desalination energy required by regenerator r.
E^{pump}_r	specific pumping energy required by regenerator r.
Q^{dil}_r	Final diluate stream flowrate of regenerator, r
$Q^{d}_{r,i}$	Diluate flowrate to sink, i
$Q^{c}_{r,i}$	Concentrate flowrate to sink, i
$Q^{k}_{j,r}$	Flowrate from source, j to regenerator, r
$Q^{s}_{j,i}$	Flowrate from source, j to sink, i
$Q^{x}_{r,r'}$	Recycle from diluate stream to regenerator feed
$Q^{y}_{r,r'}$	Recycle from concentrate stream to regenerator feed
FW	Freshwater flowrate
WW	Wastewater flowrate

Binary variable

$$y^{ED}_r = \begin{cases} 1 \leftarrow \text{if regenerator } r \text{ exists} \\ 0 \leftarrow \text{otherwise} \end{cases}$$

$$y^{d}_{r,i} = \begin{cases} 1 \leftarrow \text{piping exists between regenerator diluate and sink, } i \\ 0 \leftarrow \text{otherwise} \end{cases}$$

$$y^{c}_{r,i} = \begin{cases} 1 \leftarrow \text{piping exists between regenerator concentrate and sink, } i \\ 0 \leftarrow \text{otherwise} \end{cases}$$

$$y^{x}_{r,r'} = \begin{cases} 1 \leftarrow \text{recycle exists from regenerator diluate to regenerator, } r \\ 0 \leftarrow \text{otherwise} \end{cases}$$

$$y^{y}_{r,r'} = \begin{cases} 1 \leftarrow \text{recycle exists from regenerator concentrate to regenerator } r \\ 0 \leftarrow \text{otherwise} \end{cases}$$

$$y^{k}_{j,r} = \begin{cases} 1 \leftarrow \text{piping exists between source, } j \text{ and regenerator, } r \\ 0 \leftarrow \text{otherwise} \end{cases}$$

$$y^{s}_{j,i} = \begin{cases} 1 \leftarrow \text{piping exists between source } j \text{ and sink, } i \end{cases}$$

0←otherwise

7. References

[1] M. M. El-Halwagi, Sustainable Design Through Process Integration: Introduction to Sustainability, Sustainable Design, and Process Intergration, Waltham, MA: Elsevier Inc., 2012.

[2] R. R. Tan, D. K. S. Ng, D. C. Y. Foo and K. B. Aviso, "A superstructure model for the synthesis of single-contaminant water networks with partitioning regenerators," *Process Safety and Environmental Protection,* vol. 87, pp. 197-205, 2009.

[3] G. P. Rangaiah and T. K. Wei, Eds., Advances in Process Systems Engineering: STOCHASTIC GLOBAL OPTIMIZATION, Techniques and Applications in Chemical Engineering, vol. 2, Toh Tuck, Singapore: World Scientific Publishing Co. Pte. Ltd., 2010.

[4] M. El-Halwagi, "Synthesis of Reverse-Osmosi Networks for Waste Reduction," *AIChE Journal,* vol. 38, no. 8, pp. 1185-1198, 1992.

[5] B. Galan and I. E. Grossmann, "Optimization Strategies for the Design and Synthesi of Distributed Wastewater," *Computers and Chemical Engineering,* pp. S161-S164, 1999.

[6] C. S. Khor, D. C. Foo, M. M. El-Halwagi and Tan R. R., Shah N., "A Superstructure Optimization Approach for Membrane Separation-Based Water Regeneration Network Synthesis with Detailed Nonlinear Mechanistic Reverse Osmosi Model," *American Chemical Society,* no. 50, pp. 13444-13456, 2011.

[7] L. Yang, R. Salcedo-Diaz and I. Grossmann, "Water Network Optimization with Wastewater Regeneration Models," *Industrial & Engineering Chemistry Research,* vol. 53, pp. 17680-17695, 22 September 2014.

[8] H. Strathmann, "Electrodialysis, a mature technology with a multitude of new applications," *Desalination,* vol. 264, pp. 268-288, 2010.

[9] P. Tsiakis and L. G. Papageorgiou, "Optimal design of an electrodialysis brackish water desalination plant," *Desalination,* vol. 173, pp. 173-186, 1995.

[10] I. M. L. Chew, Tan R., Ng D. K. S, Foo D. C., Majozi T. and Gouws J., "Synthesis of Direct and Indirect Interplant Water Network," *American Chemical Society,* vol. 47, pp. 9485-9496, 2008.

AN EFFECTIVE SYNTHESIS FRAMEWORK FOR INTEGRATED WATER AND MEMBRANE NETWORKS

Esther Buabeng-Baidoo, University of the Witswatersrand, Johannesburg, South Africa

Thokozani Majozi, University of the Witswatersrand, Johannesburg, South Africa*

Abstract

Strict environmental regulations and social pressures have created the need for water and energy minimization in the process industries. This work therefore looks at the incorporation of a detailed reverse osmosis network (RON) superstructure within a water network superstructure in order to simultaneously minimize water, energy, operation and capital costs. The water network consists of water sources, water sinks and RO units for the partial treatment of the contaminated water. An overall mixed-integer nonlinear programming (MINLP) framework is developed that simultaneously evaluates both water recycle/reuse and regeneration reuse/recycle opportunities. The solution obtained from optimization provides the optimal connections between various units in the network arrangement, size and types of RO units, booster pumps as well as energy recovery turbines. The paper looks at four cases in order to highlight the importance of including a detailed regeneration network within the water network instead of the traditional 'black box' model. The importance of using a variable removal ratio in the model is also highlighted by applying the work to a literature case study which leads to a 28 % reduction in freshwater consumption and 80 % reduction in wastewater generation.

Keywords: Sustainable, Superstructure, Optimization, Synthesis, Reverse osmosis

1. Introduction

The scarcity of water and strict environmental regulations have made sustainable engineering a prime concern in the process and manufacturing industries [1]. Water minimization involves the reduction of freshwater use and effluent discharge in chemical plants. This is achieved through water reuse, water recycle and water regeneration. Water reuse involves the use of wastewater in other operations except the process where it was originally used. Water recycle, however, allows the effluent to be used in any process including the process in which it was produced. In water regeneration, the effluent is partially treated before it is recycled or reused in other processes. Partial treatment can be achieved by using water purification units often classified as membrane and non-membrane processes, e.g. RO membranes and steam stripping respectively [2].

The purification of water through membrane systems is an energy intensive process. The minimization of energy within the water networks is also needed for sustainable development. Energy usage within the water network is largely associated with the regeneration units (membrane units). In most published work, however, membrane systems

* Corresponding author: Tel: (+27) 11 7177567/7384
E-mail address: thokozani.majozi@wits.ac.za

have been represented using the "black box" approach [3-5] which uses a simplified linear model to represent the membrane systems [5]. This approach does not give an accurate representation of the energy consumption and associated costs of the membrane systems. A more rigorous representation of the regeneration unit is therefore needed [6].

The regenerator unit considered in this paper is the reverse osmosis membrane (RO). RO membranes are highly favored amongst other separation units due to their low energy consumption, ease of operation, high product recovery and quality [7]· RO membranes separate a water stream into a lean stream of low contaminant concentration known as permeate and a highly contaminated stream known as the retentate stream.

The rigorous design of RO units has been extensively studied in literature. Following on from Evangelista [8], El-Halwagi [7]introduced the 'state space approach' for the design of RO networks (RON). A superstructure representation of the RON was introduced with the objective of synthesizing a network with the optimal number of RO units, booster pumps and energy recovery turbines that resulted in minimal freshwater consumption.

There are two major approaches adopted in addressing water network synthesis namely, insight based techniques and mathematical model-based optimization methods. Insight-based techniques involve the water pinch analysis, which is a graphical method based on the concept of a limiting water profile which is the most contaminated water that can be fed into a particular operation. This method was first proposed by Wang and Smith [12]. Kuo and Smith [13] then expanded on the theory proposing a methodology for the design of treatment systems that consider both single and multiple contaminants. The computation burden of this method is, however, lower than that experienced by mathematical model-based optimization methods.

Mathematical optimization approach employs a superstructure which identifies an optimal configuration for the process from a number of alternatives. This idea was first proposed in the work by Takama *et al.* [16]. They proposed a nonlinear model that incorporates both water using and wastewater treating units for multiple contaminant systems. A significant amount of research has been conducted since then, exploring different techniques for modelling regenerators and developing solution strategies for the complex MINLPs [3, 5]. This method is however computationally expensive.

Khor *et al.* [2] presented a detailed membrane regenerator model which was incorporated into a water network superstructure (WNS). The work of Khor *et al*· [2] assumed a single regenerator with a fixed design, which implies that number of regenerators needed, number of pumps, number of energy recovery turbines were specified a priori.

The current work proposes a superstructure optimization for the synthesis of a detailed RON within a WNS. The overall mathematical formulation is used to minimize both water and energy simultaneously. A rigorous nonlinear RON superstructure model, which is based on the state space approach by El-Halwagi [7] is included in the WNS to determine the optimum number of RO units, pumps, turbines required for an optimal WNS. A fixed flowrate model that considers the concept of sources and sinks is adopted. The model takes into account streams with multiple contaminants. The idea of using variable removal ratio to describe the performance of the regenerators is also explored in this work.

2. Problem Statement

The problem addressed in this work can be stated as follows:

Given:

(i) A set of water sources, I, with known flowrates and known contaminant concentration. A set of water sinks, J, with known flowrates and known maximum allowable contaminant concentration.

(ii) A network of RO regenerators, Q, with known liquid recovery and design parameters.

(iii) A freshwater source, FW, with known contaminant concentration and variable flowrate.

(iv) A wastewater sink, WW, with known maximum allowable contaminant concentration and variable flowrate.

Determine:

(i) The minimal freshwater intake, wastewater generation, the energy consumed in the RON and the TAC.

(ii) The optimal configuration of the water network.

(iii) The optimal number of RO units, pumps and energy recovery turbines.

(iv) The optimal operation and design conditions of the RON such as feed pressure, number of hollow fiber modules per regenerator, stream distributions, separation levels etc.

3. Mathematical Model

The overall MINLP model is based on the superstructure represented in Figure 1, which is adapted from the work of El-Halwagi [7] and Khor *et al.* [2] . Figure 1 shows the incorporation of the RON superstructure with the WNS.

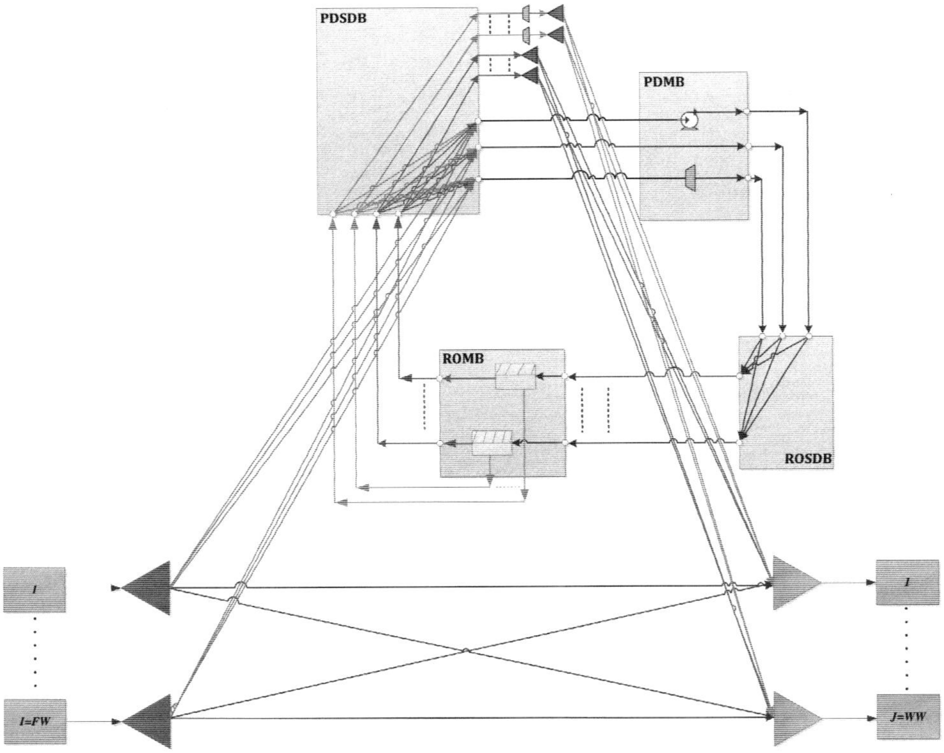

Figure 1. Superstructure representation of the RON superstructure within the WNS

3.1. Water balances

In order to ensure connectivity between the RON and the WNS, water balances are established based on the combined superstructure presented in Figure 1. From the diagram it can be seen that a water source can be fed to the RON, wastewater sink or to the water sinks. Figure 1 also shows that the water sinks receive water from the water sources, permeate and retenetate of the regeneration units as well as the freshwater source. In order to determine the existence of piping interconnections, logical constraints and discrete variables will be adopted.

3.2. Objective function

The objective function of the combined RON superstructure and WNS is used to minimize the overall cost of the regeneration network on an annualized basis which consists of:

(i). TAC of the RON

(ii). cost of freshwater (*FW)*

(iii). treatment cost of wastewater (*WW*)

(iv). capital and operation costs of the piping interconnection

The total annualized cost of the RON consists of the capital cost of RO modules, pump, and energy recovery turbines, operating cost of pumps and turbines as well as pretreatment of chemicals. The operating revenue of the energy recovery turbine is also considered in the determination of the TAC and is shown in constraint (1).

$$
\begin{aligned}
TAC(q,n) = &+C^{pump}\left(\sum_{n=1}^{N}(P_n^i - P_n^a)F_n^a\right)^{0.65} + C^{tur}\left(\sum_{n=1}^{N}(P_n^i - P_n^o)F_n^a\right)^{0.43} \qquad \forall q \in Q \quad \text{[Eq.1]}\\
&+C^{tur}\left(\sum_{j=1}^{J}(P_q^r - P_j^r)F_{q,j}^r\right)^{0.43} + C^{elec}AOT\left(\frac{\sum_{n=1}^{N}(P_n^i - P_n^a)F_n^a}{\eta_{pump}}\right)\\
&-C^{elec}AOT\left(\sum_{j=1}^{J}(P_q^r - P_j^r)F_{q,j}^r\right)\eta_{turbine} - C^{elec}AOT\left(\sum_{j=1}^{J}(P_n^i - P_n^o)F_n^a\right)\eta_{turbine}\\
&+C^{\mathrm{mod}}\sum_{q=1}^{Q}N_q^s + C^{chem}AOT\sum_{i=1}^{I}F_{i,n}^d
\end{aligned}
$$

Constraint (2) represents the objective function of the total regeneration network. It is also assumed that all the pipes share the same properties of *pc* and *qc* and a 1-norm distance D.

$$\min \left(\begin{array}{l} \sum_{q=1}^{Q}\sum_{n=1}^{N} TAC_{q,n} + AOTC^{water} FW + AOTC^{waste} WW \\ + AA\left(\sum_{i=1}^{I}\sum_{j=1}^{J} D_{i,j}\left(\frac{p_c F_{i,j}^{s}}{3600v} + q_c y_{i,j}\right)\right) \\ + AA\left(\sum_{q=1}^{q}\sum_{j=1}^{J} D_{q,j}^{p}\left(\frac{p_c F_{q,j}^{p}}{3600v} + q_c y_{q,j}^{p}\right)\right) \\ + AA\left(\sum_{q=1}^{q}\sum_{j=1}^{J} D_{q,j}^{r}\left(\frac{p_c F_{q,j}^{r}}{3600v} + q_c y_{q,j}^{r}\right)\right) \\ + AA\left(\sum_{i=1}^{I}\sum_{n=1}^{N} D_{i,n}^{d}\left(\frac{p_c F_{i,n}^{d}}{3600v} + q_c y_{i,n}^{d})\right)\right) \end{array} \right) \quad [Eq.2]$$

The overall model results in a nonconvex MINLP due to the bilinear terms as well as the power function in the constraints.

4. Illustrative example

The above model was applied to a petroleum refinery case study based on the work presented by Khor *et al.* [2]. The model was implemented in GAMS 24.2 using the general purpose global optimization solver BARON which obtains a solution by using a branch-and-reduce algorithm. The network consists of 4 sources and 4 sinks. The limiting water data for the sources and sinks is given in Table 1.

Table 1. Limiting data for water network

Sources, *i*				**Sinks, *j***			
i	**Flowrate (kg/s)**	**Contaminant Concentration (kg/m^3)**		***j***	**Flowrate (kg/s)**	**Max Contaminant Concentration (kg/m^3)**	
		TSS	COD			TSS	COD
1	7.3	3.5	3.5	1	0.83	2.5	2.5
2	10.65	4	4	2	40	2	2
3	3.5	1	3	3	5.56	2.5	2.5
FW	—	2	1	*WW*	—	25	25

Table 2 shows the comparison between a case where multiple regenerators with fixed (case 1) and variable removal ratio (Case 2) where used. The removal ratio chosen by the model in case 2 was 0.97 for all contaminants instead the fixed value of 0.95 in case 1. Case 2 led to 3.12 % reduction in freshwater and 30.43 % reduction in wastewater generation in comparison with case 1. A 15.91 % reduction in the total network cost was also achieved. The large decrease in the total cost of the network in case 2 can be attributed to the high removal ratio which was selected by the model than the value that was initially predicted. The modeling of case 2 is however computationally expensive as can be seen in Table 2. Figure 2 shows the water network for case 2 with the corresponding flowrate for each stream. The best case used 15 HFRO modules per regenerator. The model selected two regenerators, two pumps and two energy recovery turbines as can be seen in Figure 2. It can also be seen that a parallel configuration of the network was chosen by the model. In comparison with the case where no regeneration was considered, case 2 leads to a 28 % reduction in freshwater consumption and 80 % reduction in wastewater generation.

Table 2. Summary of results for case 1 and 2

	Multiple regenerators	
	Fixed RR (Case 1)	Variable RR (Case 2)
Freshwater flowrate (kg/s)	28.87	27.68
Wastewater flowrate (kg/s)	3.91	2.72
Cost of regeneration (million $/year)	0.23	0.096
Total cost (million $/year)	1.32	1.11
Network configuration	Parallel	Parallel
Number of HFRO modules	37 for each regenerator	15 for each regenerator
CPU time (h)	6	54

The high computational time for solving the model in case 2 was due to the complexity of the problem as well as the large number of 0-1 variables. The model solves quicker when tighter bounds are imposed on the feed and retentate pressure. The use of the energy recovery turbines in the RON led to a reduction in the regeneration cost of the network, and as a result, a reduction in energy usage by the system was achieved.

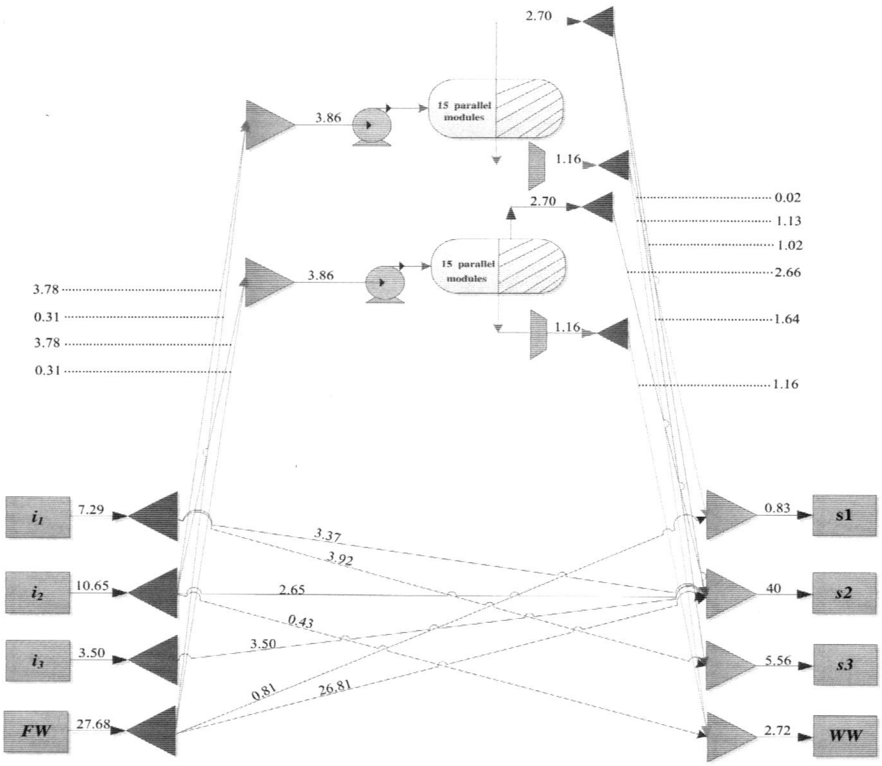

Figure 2. Network obtained for case 2 (multiple regenerators with variable removal ratio)

5. Conclusion

This paper has addressed the synthesis of a water regeneration network that incorporates the detailed synthesis of an RON. The proposed model was applied to a literature case study and was then solved using GAMS/BARON in order to highlight its practicality. The results show that the use of multiple regenerators in the water network, can lead to a reduction in the total cost of the network due to the significant reduction in freshwater consumption and wastewater generation. It can also be concluded that, there is a significant benefit in allowing the removal ratio in the model to be a variable as this has severe significant on the cost and structure of the network. Large computational times were however incurred due to the complex nature and structure of the model. It is also noteworthy that the proposed model was limited to one membrane technology. Multiple membrane technologies such as ultrafiltration can however be incorporated in the membrane network and thus offering a scope for future work.

6. Nomenclature

6.1. Sets

$I=\{i|i=\text{water source}\}$

$J=\{j|j=\text{water sink}\}$

$M=\{m|m=\text{contaminants}\}$

$Q=\{q|q=\text{regeneration units}\}$

6.2. Parameters

AOT	annual operating time
p_c	parameter for carbon steel piping based on CEPCI value of 318.3
q_c	parameter for carbon steel piping based on CEPCI value of 318.
v	velocity
η_{pump}	pump efficiency
$\eta_{turbine}$	turbine efficiency
$D^a_{i,j}$	manhattan distance between water source i and sink j
$D^p_{q,j}$	manhattan distance between regenerator q and sink j
$D^r_{q,j}$	manhattan distance between regenerator q and sink j
$D^d_{i,n}$	manhattan distance between source i and node n
C^{chem}	cost parameter for chemicals
C^{elec}	cost of electricity
C^{mod}	cost per module of HFRO membrane
C^{pump}	cost coefficient for pump
C^{tur}	cost coefficient for turbine
P^r_j	pressure of the retentate stream in sink j
AA	annualisation factor

6.3. Continuous Variables

$F^s_{i,j}$	allocated flowrate between sources i and sinks j
$F^d_{i,n}$	allocated flowrate between sources i and node n
$F^p_{q,j}$	flowrate of the permeate stream from regenerators q to sinks j
$F^r_{q,j}$	flowrate of the retentate stream from regenerators q to sinks j
F^a_n	flowrate of streams from re node n
P^a_n	pressure of streams leaving node n
P^i_n	pressure of an inlet stream to an energy recovery turbine from node n
P^o_n	pressure of an outlet stream from an energy recovery turbine from node n
P^r_q	pressure of a retentate stream from regenerator q
FW	freshwater flowrate
WW	wastewater flowrate

6.4. Binary Variables

$$y^{p}_{q,j} = \begin{cases} 1 \leftarrow \text{if piping exits between the permeate streams and sink } j \\ 0 \leftarrow \text{otherwise} \end{cases}$$

$$y^{r}_{q,j} = \begin{cases} 1 \leftarrow \text{if piping exits between the retentate streams and sink } j \\ 0 \leftarrow \text{otherwise} \end{cases}$$

$$y_{i,j} \begin{cases} 1 \leftarrow \text{if piping exits between source } i \text{ and sink } j \\ 0 \leftarrow \text{otherwise} \end{cases}$$

$$y^{d}_{i,n} \begin{cases} 1 \leftarrow \text{if piping exits between source } i \text{ and node } n \\ 0 \leftarrow \text{otherwise} \end{cases}$$

6.5. Integer Variables

N^{s}_{q} the number of hollow fiber modules of regenerator q

7. References

[1] S. Bandyopadhyay and C. Cormos, "Water management in process industries incorporating regeneration and recycle through a single treatment unit," *Industrial and Engineering Chemistry Research* , vol. 47, pp. 1111-1119, 2008.

[2] C. Khor, D. Foo, M. El-Halwagi, R. Tan and N. Shah, "A superstructure optimization approach for membrane separation-based water regeneration networks synthesis with detailed nonlinear mechanistic reverse osmosis model," *Industrial and Engineering Chemistry Research* , vol. 50, pp. 13444-13456, 2011.

[3] A. Alva-Argáez, A. Kokossis and R. Smith , "Wastewater minimisation of industrial systems using an integrated approach," *Computers and Chemical Engineering,* vol. 22, pp. 741-744, 1998.

[4] C. Khor, B. Chachuat and N. Shah, "A superstructure optimization approach for water network synthesis with membrane separation-based regenerators," *Computers and Chemical Engineering* , vol. 42, pp. 48-63, 2012 .

[5] R. Tan, D. Ng, D. Foo and K. Aviso, "A superstructure model for the sybthesis of single-contaminant water network with partitioning regenerators," *Process Safety and Environmental Protection,* vol. 87, no. 3, pp. 197-205, 2009.

[6] C. Khor , B. Chachuat and N. Shah, "Optimization of water network synthesis for single-site and continous processes: milstones, challenges, and future directions," *Industrial & Engineering Chemistry Research* , vol. 53, pp. 10257-10275, 2014.

[7] M. El-Halwagi, "Synthesis of reverse osmosis networks for waste reduction," *American Institute of Chemical Engineers,* pp. 1185-1198, 1992.

[8] F. Evangelista, "A short-cut method for the design of reverse-osmosis desalination plants," *Industrial & Engineering Chemistry Process Design and Development,* vol. 24, p. 211, 1985.

[9] Y. Wang and R. Smith, "Wastewater minimization," *Chemical Engineering Science* , vol. 49, no. 7, pp. 981-1006, 1994.

[10] W. Kuo and R. Smith, "Effluent treatment system design," *Chemical Engineering Science* , vol. 52, pp. 4273-4290, 1997.

[11] N. Takama, T. Kuriyama , K. Shiroko and T. Umeda , "Optimal water allocation in a petroleum refinery," *Computers and Chemical Engineering,* vol. 4, pp. 251-258, 1980.

Effect of Grid Generation in Numerical Simulation of Spray Granulation Tower

Liuhai Feng, Jiangyun Wang, Juan Wang, Fan Zhao, Yu Mao*
College of Chemical Engineering, China University of Petroleum, Beijing, P.R. China.

Abstract

The nozzle is the core equipment in the spray granulation tower. The separation efficiency of the Tower can be improved by adjusting the swirl intensity. However, the declining nozzle in the spray granulation tower increases the difficulty of geometric modeling. Researchers simplified the structure by establishing a vertical jet model with specific boundary conditions. To ensure the same flow rate with no-simplified nozzle, the researchers often increased the jet velocity or jet area, which often led to a deviation between the simulation result and actual result. To compare the different simulation result between the simplified nozzle and no-simplified nozzle, this paper researched the flow distribution in the spray granulation tower with arraying nozzles through the three-dimensioned numerical by numerical simulation and experiments respectively. The comparison of simulation and experimental result indicates that flow behavior and velocity distribution are difference between different grids. There is a longer central jet area and smaller atomization angle in no-simplified nozzle structure, which can weaken the numerical dissipation resulted from the inconsistency between flow direction and gridding direction near the nozzle entry. The simulation result of no-simplified spray granulation tower agrees well with experimental result. Therefore, it can obtain flow feature more accurately using no-simplified nozzle spray granulation tower.

Introduction

Due to the decreasing reserves and increasing weight of crude oil, the processing of heavy oil can effectively alleviate the serious energy crisis. Coupled with serious environmental problem, the process technology of heavy oil is obliged to be upgraded with higher efficiency. Heavy oil deep stage separation[1, 2] is a technique using supercritical solvent to extract asphalt which is hard to deal with in heavy oil. The processed oil is the important feedstock component of hydro-treating. The extracted materials are separated of solid solute from gas solvent in spray granulation tower. To improve the separation efficiency, it is necessary to study the flow behavior in the spray granulation tower. During the industrial oil production process, the fluid in the spray tower is the superficial fluid with drastic flash transition, so the flow process in the tower is extremely complicated. The experimental measurement is conducted under the atmospheric pressure and ambient temperature conditions, illustrated in Fig1.

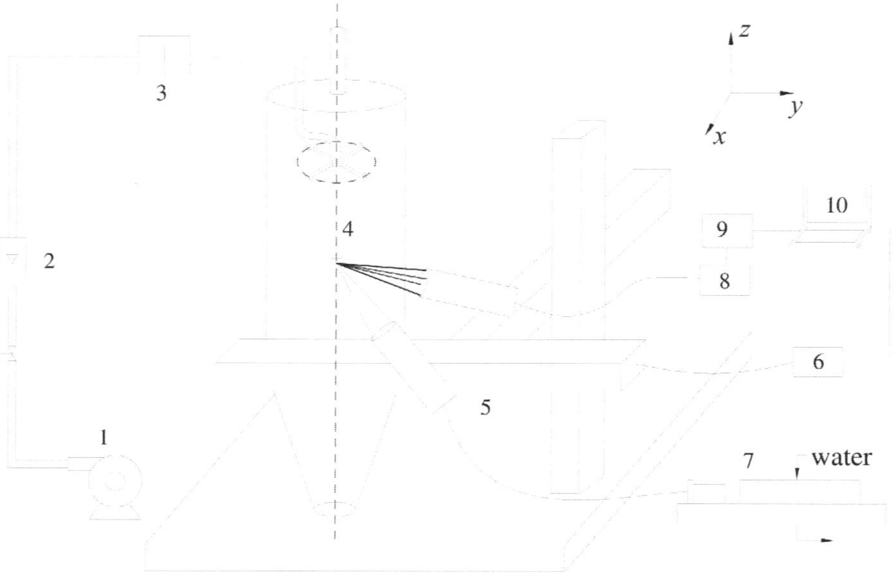

(1) compressor; (2) rotameter; (3) tracer particle generator; (4) spray granulation tower; (5) 3-D traverse system; (6) coordinate manipulator; (7) laser; (8) converter; (9) signal processor; (10) computer.

Fig.1 Schematic diagram of spray granulation tower

Simulation condition

The declining nozzle in the spray granulation tower increases the difficulty of geometric modeling. Langrish[3] and Liu[4] simplified the declining nozzle by establishing a vertical jet model with setting specific boundary conditions, which is used to ensure the consistence of the jet angle and the rate of flow between simulation and experiment.

According to the formula:

$$\begin{aligned} Q &= Av && \cdots No-simplified \\ &= \frac{A}{\cos\alpha} v\cos\alpha && \cdots Variation\ on\ orifice\ area \\ &= A\frac{v}{\cos\alpha}\cos\alpha && \cdots Variation\ on\ jet\ velocity \end{aligned} \qquad (1)$$

so the jet area or velocity should be increased by 1/cosα times in simplified nozzle , respectively.

A structured hexahedron grid is used to establish the three different CFD models. The major difference of the three models is the spatial position of the nozzle. The flow in the tower is defined as turbulence according the Reynolds number. There is a rotational flow field in the tower, so the anisotropic RSM model[5, 6] is chosen as the calculating model. The nozzles are defined as Velocity Inlets, where the velocity magnitude and the flux are specified according to the experiment. The flow at the outlet of the tower is fully developed, so an ambient pressure is given at the outlet. The walls are set as the no-slip boundary conditions, and the standard wall function is applied to determine the flow near walls. The computation is carried out by the commercial computer software, Fluent. The governing equations is discretized by control volume integral method, and the SIMPLE algorithm is implemented to deal with the coupling between pressure and velocity, The solution is considered to be converged when the residual of the continuity is less than 10^{-4}.

Result and discussion

In consideration of the influence caused by grid generation, the spray granulation tower with single nozzle is simulated first in this study. The angle between the nozzle and the vertical plane is 45°, the diameter of the nozzle is 4 mm, and the jet velocity is 10 m/s. The cloud chart of velocity distributions is shown in Fig.2. Comparing the red elliptical benchmark, it finds that the atomizing angle of the no-simplified nozzle is obviously smaller than that of in simplified nozzle. So the increase of jet area or the jet velocity can enlarge the angle of atomization, and the increase of velocity has a more obvious influence on the change of angle. Comparing the black rectangular benchmark, it finds that the jet zone of no-simplified nozzle is longer than that of the simplified nozzles, and the increase of jet area has obvious influence on the attenuation of velocity. The central velocity along the jet direction is shown in Fig3. In the three models, the velocity has a same trend when $S > 0.08$ m. The air velocity of no-simplified nozzle maintains stability with some distance and decays subsequently. However, the air velocity of simplified nozzle decays quickly. In conclusion, the simplified structure has a great influence on the simulation result.

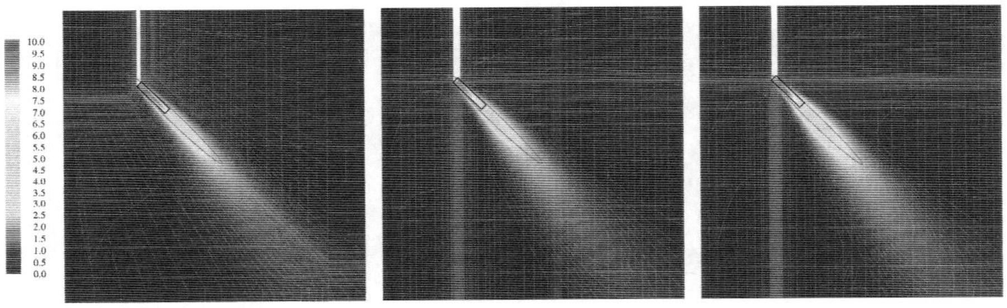

(a) No-simplified; (b) Variation on orifice area; (c) Variation on jet velocity

Fig.2 Comparison of different grid models

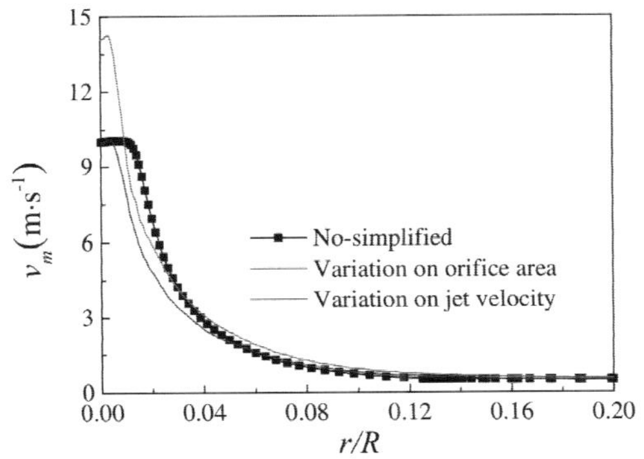

Fig.3 Variation of velocity along jet direction

Below the nozzle, the flow is influenced obviously by jet and the flow field is chaotic. The flow forms a stable rotational flow field as time goes on. So the flow field in the spray tower is divided into two parts: jet-dominated area and vortex stability region. The research in this article is mainly focus on the stability region, as shown in Fig.4. It can be seen that both the forced vortex in center and the free vortex outside can be acquired from simulation result in every structure. But the velocity distribution trend got from simulation result of simplified nozzle has some differences with the experimental result.

However, the simulation result of no-simplified nozzle coincides with experimental result better, including the peak value and trend of the tangential velocity. In conclusion, the no-simplified nozzle describes the flow process in the spray tower more accurately.

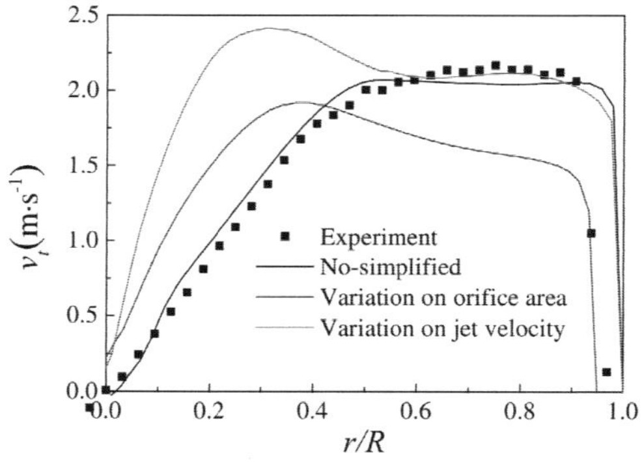

Fig.4 Comparison of velocity in simulation and experimental result

The simulation result is analyzed which is obtained from no-simplified nozzle spray granulation tower. Due to the interaction of jet and swirl flow, the flow field formed in SGT is complex, as shown in Fig.5. The flow in jet-dominated zone is greatly influenced by the jets. The high jet gas entrains the surrounding medium to the wall, and a spiral downward flow is formed as a result of the restriction of the wall. Besides, there are a lot of secondary vortices which the direction is oppose to the mainstream near the wall. The vertical eddy and secondary vortices are formed near the wall due to the combined effect of the jet and the entrainment. The flow pattern in the stable swirling zone is good. The effect of jet weakens gradually, resulting in the formation of the stable Rankine vortex swirl flow. The stable swirling zone is divided into two regions: an outer downward flow region and the center upward flow region. There exist a radial flow from outer to the center, causing turbulence in center region.

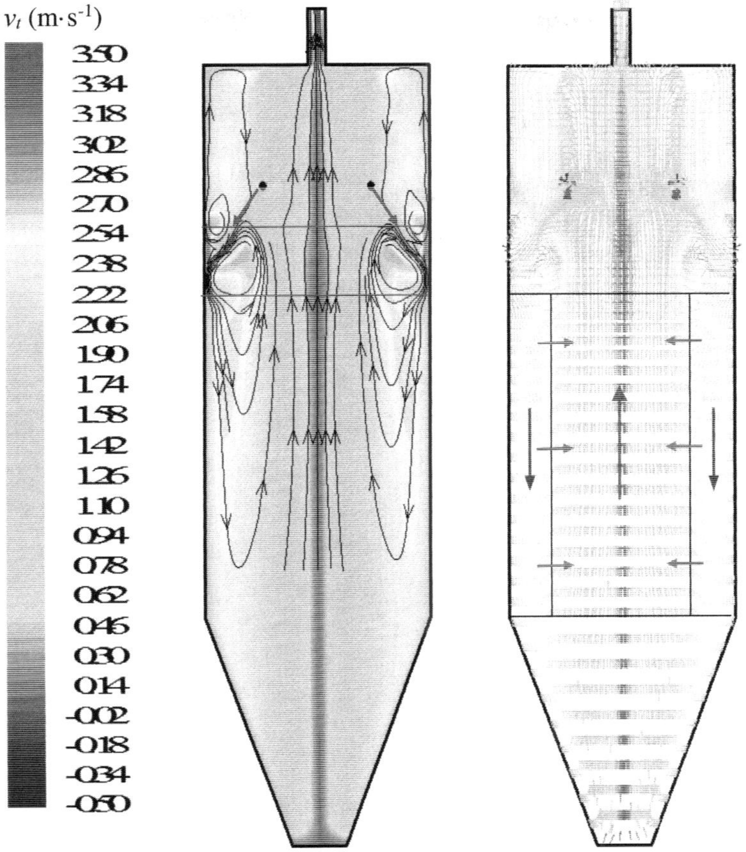

Fig.5 Distribution of tangential velocity and velocity vector in vertical section

Conclusions

The gas flow behavior in a jet-driven swirling flow is studied experimentally and numerically in this work. The flow dynamics, including flow development, velocity field and division characteristic, are revealed. The main conclusion can be summarized as follows:

(1) The no-simplified nozzle has a longer jet zone and smaller atomization angle, and the simulation results of the no-simplified nozzle has some difference with the experimental results. The simplified structure has some influence on the simulation results.

(2) Comparison of the simplified nozzle, the peak value and trend of the tangential velocity got from the no-simplified simulation coincide with the experimental result. The no-simplified structure of nozzle can weaken the numerical dissipation resulted from the inconsistency between flow direction and gridding direction near the outlet of the nozzle.

(3) There exists two regions in the spray tower: the jet-dominated area has a complex flow distribution, which has some vertical circulation and secondary vortex; the rotary area has a stable flow field, which has a flow distribution similar with the RanKine Vortex.

References

1. Shi T P, Hu Y X, Xu Z M, et al. Characterizing petroleum vacuum residue by supercritical fluid extraction and fractionation[J]. *Industrial & engineering chemistry research*, 1997, 36(9): 3988-3992.
2. Zhao S Q, Xu C M, Wang R A, et al. Deep separation method and processing system for the separation of heavy oil through granulation of coupled post-extraction asphalt residue [P]. US Patent, 7597794 B2, 2009.
3. Langrish T A G, Williams J, Fletcher D F. Simulation of the effects of inlet swirl on gas flow patterns in a pilot-scale spray dryer[J]. *Chemical Engineering Research and Design*, 2004, 82(7): 821-833.
4. Liu M, Mao Y, Wang J, et al. Effect of swirl on hydrodynamics and separation performance of a spray granulation tower with array nozzles [J]. *Powder Technology*, 2012, 227: 61-66.
5. Liu F, Chen J, Zhang A, et al. Performance and flow behavior of four identical parallel cyclones [J]. *Separation and Purification Technology*, 2014, 134: 147-157.
6. Zhao B, Su Y, Zhang J. Simulation of gas flow pattern and separation efficiency in cyclone with conventional single and spiral double inlet configuration [J]. *Chemical Engineering Research and Design*, 2006, 84(12): 1158-1165.

13 Ways Through a Firewall
What You Don't Know *Will* Hurt You

Compensating Measures

Attack Type	2FACT	ENC	RULES	HOST	NET	SUPD	UGW
1) Phishing / trojan / drive-by-download – victim pulls attack through firewall	0	0	2	1	1	1	2
2) Social engineering – steal a password / keystroke logger / shoulder surf	2	0	0	0	0	0	2
3) Compromise domain controller – create control system or firewall account	1	0	2	0	0	0	2
4) Attack exposed servers – SQL injection / DOS / buf-overflow / default pwds	0	1	1	1	1	1	2
5) Attack exposed clients – compromised web servers / file servers / data svrs	0	0	2	1	1	1	2
6) Session hijacking – MIM / steal HTTP cookies / command injection	0	2	1	0	1	0	2
7) Piggy-back on VPN – split tunneling / malware propagation	1	1	2	1	1	1	2
8) Firewall vulnerabilities – bugs / zero-days / default password / design vulns	0	0	0	0	1	1	2
9) Errors and omissions – bad firewall configurations / IT reaches through fwalls	1	1	1	1	1	1	2
10) Forge an IP address – firewall rules are IP-based	1	1	0	1	1	1	2
11) Bypass network perimeter – rogue cabling / wireless / cell phones / dial-up	1	1	0	1	1	1	0
12) Physical access to firewall – administrator ports / no password / modify hw	0	0	0	0	0	0	0
13) Sneakernet – removable media / untrusted laptops	0	0	0	1	0	1	0
Total Score:	7	7	11	8	9	9	20

Photo: Red Tiger Security

Abbr.	Compensating Measure
2FACT	2-factor authentication
ENC	Encryption, cryptographic authentication
RULES	Better firewall rules
Host	Host intrusion detection / prevention systems & SIEMs
Network	Network intrusion detection / prevention systems & SIEMs
SECUPD	Security updates / patch programs
UGW	Unidirectional security gateways

Grade	Description
2	Blocks essentially all attacks in this class
1	Blocks some attacks in this class
0	Not effective at blocking this class of attacks

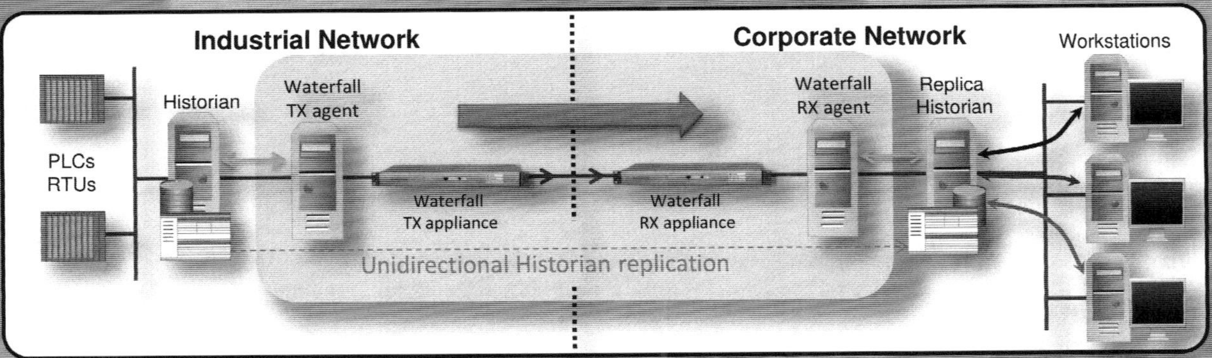

Unidirectional Security Gateways: Absolute Protection From External Network Attacks

400436 Optimisation of Refinery Diesel Blending

Abstract:

OPTIMISATION OF REFINERY DIESEL BLENDING

Shixun Jiang, Nan Zhang

Centre for process integration, University of Manchester

Oxford Road, Manchester M13 9PL, UK

In recent years, modern refineries have been confronted with higher prices of petroleum, more stringent environment standards and stricter requirements of quality specifications. To achieve the objective of maximizing the profit, it brings great challenge toward planning and scheduling for refinery operators.

Diesel, one of the main petroleum products, has a growing market proportion because of its higher energy utilization ratio and therefore lower carbon dioxide emissions compared with gasoline. Diesel is widely used in the industry and transportation. Only high quality diesel product can survive in the more and more competitive market. Refiners haven't stopped improving the methodology in diesel production and management.

To maximize the profit or minimize the cost is the main driving force in refinery diesel management. It is difficult to achieve the optimal recipe towards the site level as an optimal recipe for one single process may be suboptimal for other processes or even infeasible for the overall refinery. A multi-period planning model for refineries, which was based on the decomposition method, was proposed by Dave in 2004, to optimise decisions at different time intervals. Furthermore to optimise the recipe, refinery operators have to deal with volatile prices of crude oil and products, fluctuant market demand, uncertainty of feedstock arrivals and other abrupt changes. Sourabh (2008) demonstrated an "Operating Window" algorithm, which decomposed an MINLP problem into an NLP model and an MILP model.

In this work, a model for management of refinery diesel streams has been developed to optimize the diesel production of a refinery. A modified property estimation model for diesel, which is more accurate than traditional linear models, is employed for off-line diesel blending processes. Intricate environment regulations and product specifications, such as cetane number, sulphur content and pour point, etc. are taken into account. A novel scheduling method, which involves an MINLP model, is optimized by a robust solving algorithm. By adopting the method, the optimal solution of schedule a refinery diesel streams can be reached. Case studies have been carried out to demonstrate the effectiveness of the developed method.

References

Dhaval Dave, 2004. Planning and scheduling of refinery operations. Ph.D Thesis, University of Manchester, Faculty of Engineering and Physical Sciences.

Sourabh Gupta., 2008. Planning and scheduling of refinery operations. Ph.D. Thesis, University of Manchester, Faculty of Engineering and Physical Sciences.

400598 Electrochemical Reactor Modeling and the Determination of Minimum Operating Parameters for Conductive Paint Antifouling Systems

Abstract:

Protection from fouling organisms (biofouling) is a key function of coatings on marine hydrokinetic (MHK) devices. Accumulation of biomass on devices risks performance degradation and possible failure. Additionally, artificial structures in the marine environment have been shown to encourage settlement of invasive species, necessitating application of mitigation strategies.[1] At the present there are no proven reliable options for prevention of fouling in the marine environment with applications to match device lifetimes expected to be on the order of 10-20 years. Removal, cleaning, and re-application of short-lived antifouling paints is an expensive prospect. Electrified conductive paints held at sufficient electric potential have been shown to be effective in the prevention of marine biofouling, and may be a viable solution for long term antifouling performance[2] . The present work seeks to extend the understanding of performance variables for such coatings. Modeling is used to predict chemical species concentrations in fluid flow over a charged flat plate. Minimum operating conditions required for inhibition of algal growth in filtered seawater are then determined experimentally in a flow cell with applicable geometry to elucidate the mechanism of action.

The mechanism of action has been proposed variously to be direct electron transport[2], electromagnetic field repulsion[3], and electrochemical reaction[4, 5]. Evidence suggests that electrochemical oxidation of chloride and bromide present in seawater produces strong oxidizing agents which sanitize the positively charged anodic surface. The negatively charged cathodic surface is likely driving the hydrogen evolution reaction, which increases the local pH to intolerable levels for marine fouling organisms. Comparison of the experimentally determined minimum potential needed for biofouling protection with predicted equilibrium potentials for each reaction and Tafel plots in filtered seawater support this hypothesis.

[1] L. H. Hedge and E. L. Johnston, "Propagule pressure determines recruitment from a commercial shipping pier," *Biofouling,* vol. 28, pp. 73-85, 2012/01/01 2012.

[2] T. Matsunaga, T. Nakayama, H. Wake, M. Takahashi, M. Okochi, and N. Nakamura, "Prevention of marine biofouling using a conductive paint electrode,"*Biotechnol. Bioeng.,* vol. 59, pp. 374-378, // 1998.

[3] T. Nakayama, H. Wake, K. Ozawa, N. Nakamura, and T. Matsunaga, "Electrochemical prevention of marine biofouling on a novel titanium-nitride-coated plate formed by radio-frequency arc spraying,"*Appl. Microbiol. Biotechnol.,* vol. 50, pp. 502-508, // 1998.

[4] R. E. Perez-Roa, M. A. Anderson, D. Rittschof, C. G. Hunt, and D. R. Noguera, "Involvement of reactive oxygen species in the electrochemical inhibition of barnacle (Amphibalanus amphitrite) settlement," *Biofouling,* vol. 25, pp. 563-71, // 2009.

[5] J.-R. Huang, W.-T. Lin, R. Huang, C.-Y. Lin, and J.-K. Wu, "Marine biofouling inhibition by polyurethane conductive coatings used for fishing net," *J. Coat. Technol. Res.,* vol. 7, pp. 111-117, // 2010.

[6] *Paint and Coating Testing Manual : 15th edition of the Gardner-Sward handbook*: ASTM International, 2012.

400962 Advanced 3-Phase Flash Computations in the API Tech Data Book 9 Software

Abstract:

Flash computations and the physical properties of each phase are vitally important for accurately sizing process equipment. The 3-phase flash/mixture calculations in the API Tech Data Book 9th Edition Software provide highly accurate flash results for hydrocarbon mixtures. This is due to the incorporation of accurately tuned binary interaction parameters and the use of industry proven equation of state methods for a very wide range of components and operating conditions. This allows for accurate flash calculations to be done by any chemical engineer working in the oil and gas industry for most systems of interest without any indepth knowledge of thermodynamics and thermophysical properties. Comparison of results by the API Tech Data Committee against 3 major simulaiton software products showed superior results in 5 areas important to accurately sizing process equipment. The software interface uses a modern XAML design that does not require a flowsheet to be created so that chemical engineers not experienced with process simulation software can quickly become proficeint in its use.

CD-adapco

2015 AIChE Spring Meeting &
11th Global Congress on Process Safety
April 26 - 30, 2015 • Hilton Austin • Austin, TX
www.aiche.org/spring

A CFD Model-Based Optimization of a Process Burner Geometry

Krishnamoorthy, N., Halliday, N., Dewan, Y., and Aglave, R.,

Problem Description

CD-adapco

Design optimization of a single flat flame burner using STAR-CCM+ and Optimate+

Optimate+ is an easy to use plugin that utilizes SHERPA algorithm for design space exploration. It is:

- Hybrid – blend of search strategies, global and local search
- Adaptive – adapts to design space, searches simple and complex spaces

Objectives:

- Minimize: CO and NOx relative to the baseline design
- Explore: The dependence of Flame Height and Emissions on burner design parameters
  - Influence of tile tilt angle
  - Influence of fuel port spacing

Geometry and Operating Conditions

- 4 Fuel ports
- Fuel and oxidizer at STP
- Burner Heat Release = 0.88 MW
- Excess Air Supplied = 10%

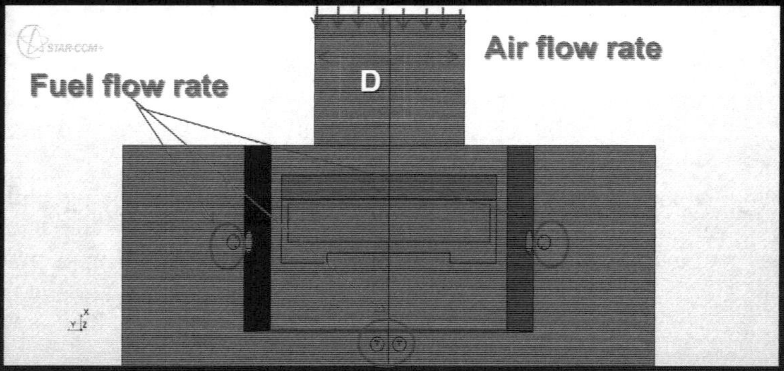

Numerical Model and Optimization Methodology

Numerical Model Selection in STAR-CCM+

- Three Dimensional
- Half Burner : Symmetry Condition
- SST K-Omega Turbulence Model
- PPDF Flamelet Combustion Model
- Thermal NOx: NOx PPDF Flamelet
- Discrete Ordinates Radiation Model
  - Weighted Sum of Gray Gases

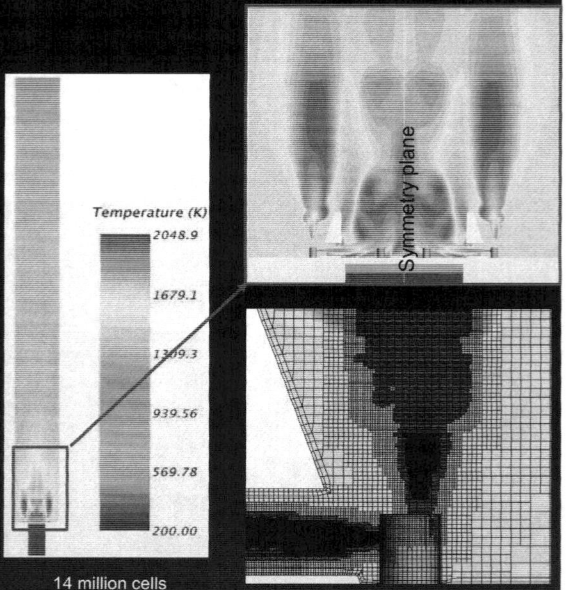

14 million cells

Fuel jet resolution using table based refinement

Optimization Variables

A: Primary Injector Spacing
Minimum: 25 mm
Maximum: 250 mm
Intervals: 50

B: Burner Tile Angle
Minimum: 1
Maximum: 35
Intervals : 10

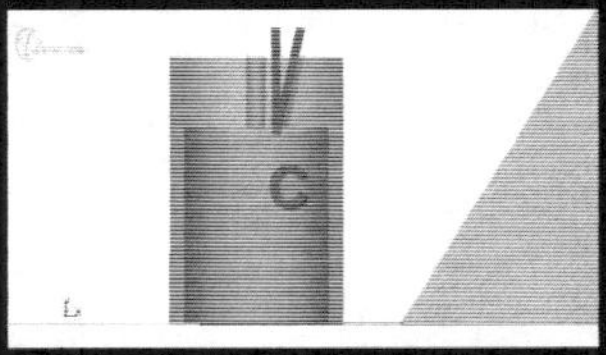

C: Primary Inj. Nozzle Angle
Minimum: 1
Maximum: 20
Intervals : 10

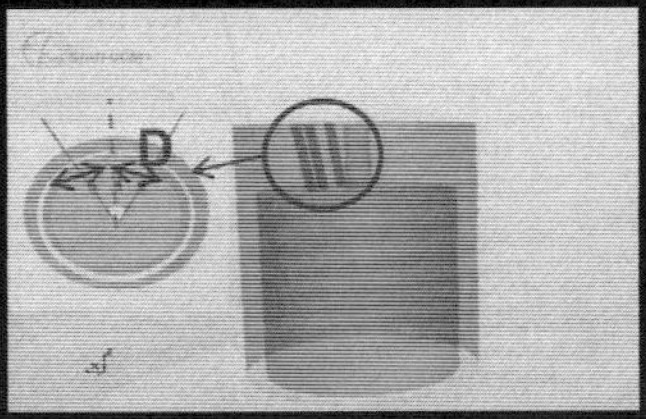

D: Primary ports angle spacing in circular pattern
Minimum: 5
Maximum: 45
Intervals : 11

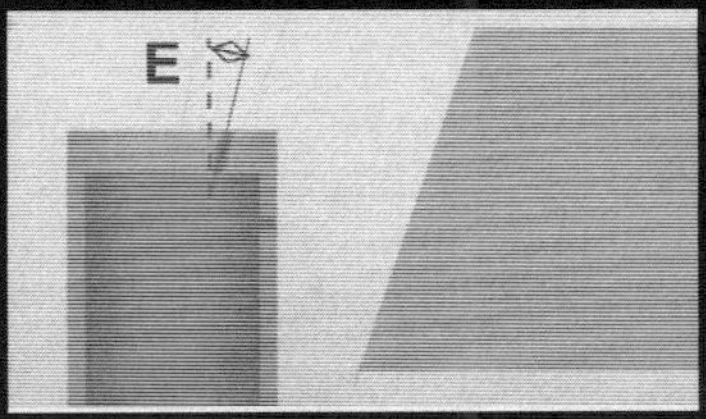

E: Side Inj. nozzle angle
Minimum: 1
Maximum: 30
Intervals : 10

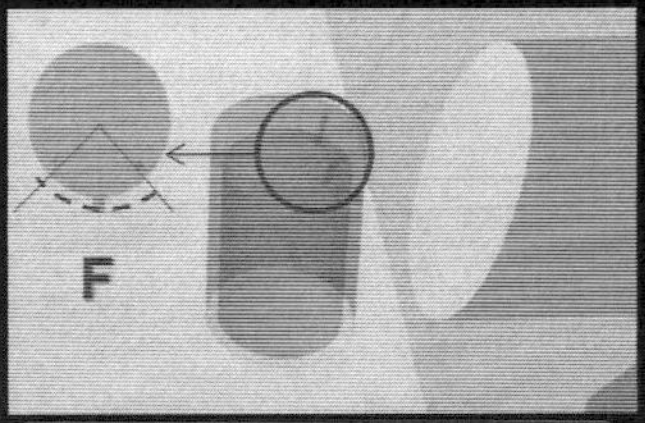

F: Side nozzle angle spacing in circular pattern
Minimum: -120
Maximum: 60
Intervals : 50

Burner Optimization: Responses and Constraints

Responses

- **Flame Height**
  - Defined by iso-contour of CO = 1200 ppm
- **Flame Volume**
  - Volume occupied by iso-contour of CO = 1200 ppm
- **CO concentration at Exit**
  - Mass flow averaged CO mass fraction at outlet
- **NOx concentration at Exit**
  - Mass flow averaged NOx mass fraction at outlet
- **Temperature at Exit**
  - Mass flow averaged Temperature at outlet

Constraints

- **Volume Average Temperature**
  - In the firebox > 1350 K
- **Flame height**
  - > 1.695 m (30% of baseline)

Pareto Optimization Results

CD-adapco

- **Minimize CO and NOx emissions at exit**
  - CO and NOx are competing objectives

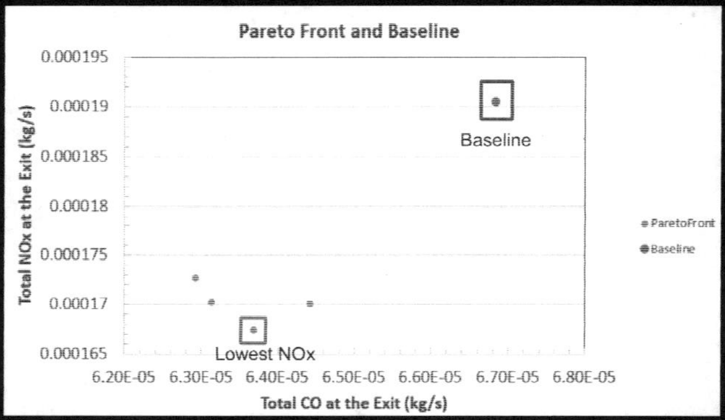

- **Total of 40 designs were run**
- **15 designs did not meet the constraint. These were rejected**

Best design minimizes CO by 4.7% and NOx by 12% from the baseline

Influence of Tile Tilt Angle

CD-adapco

Design 1 | **Design 13** | **Design 32**

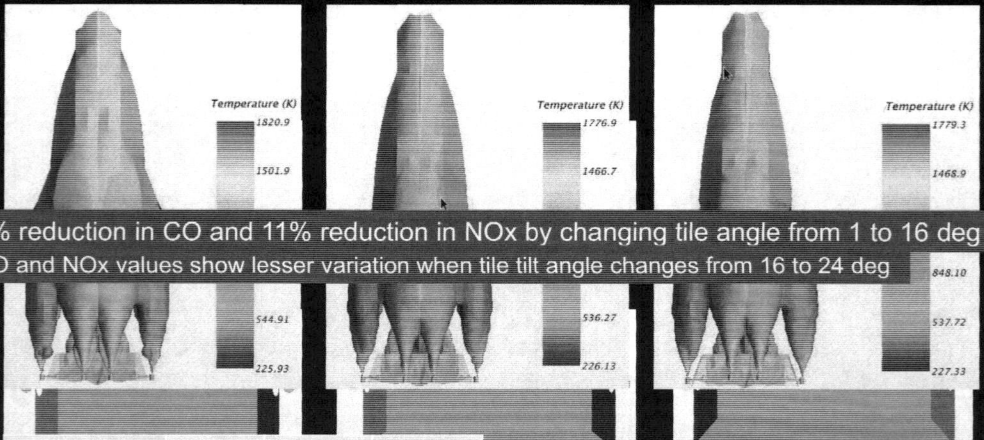

4% reduction in CO and 11% reduction in NOx by changing tile angle from 1 to 16 deg

CO and NOx values show lesser variation when tile tilt angle changes from 16 to 24 deg

	Design 1	Design 13	Design 32
Flame Height (m)	2.422207	2.342207	2.42221E+00
Flame Volume (m^3)	2.15E-01	1.86E-01	1.84E-01

3% reduction in flame height for Design 13

Similar flame height to baseline for Design 32

~14% reduction in flame volume for Designs 13 and 32

Influence of Fuel Injector Spacing

CD-adapco

Design 17
Design 17
Injector Spacing 25 mm

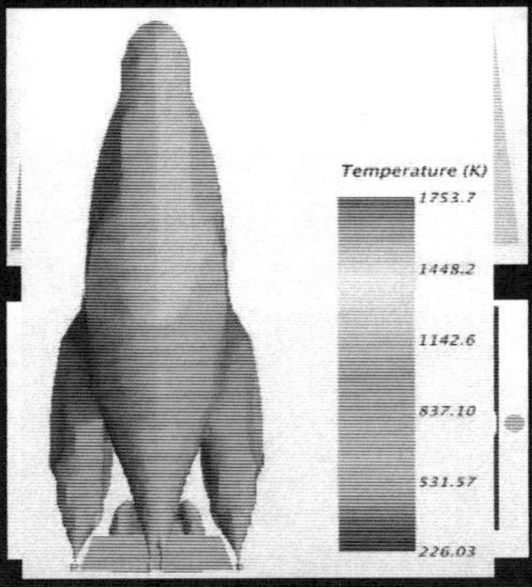

Design 1
Design 1
Injector Spacing 88 mm

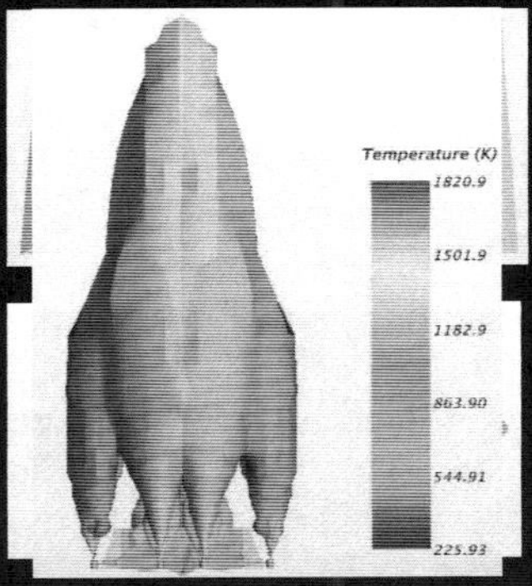

Design 28
Design 28
Injector Spacing 250 mm

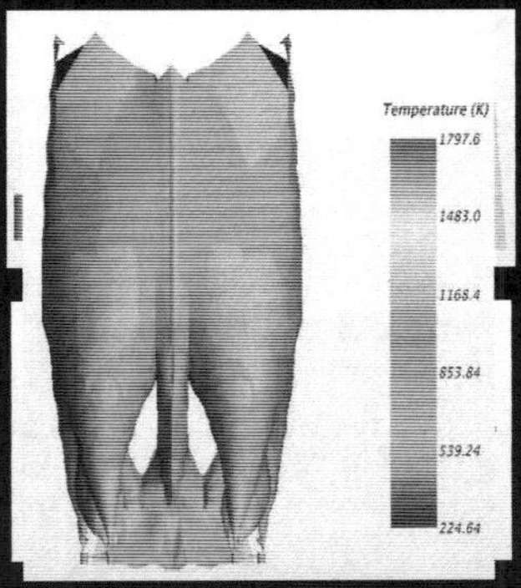

Flame height and Flame volume for Design17 higher than Design 1. Design 17 has 8% higher NOx

5% reduction in CO and 3.4% reduction in NOx for Design 28. But Flame height reduced by 28%

Summary of Design Optimization

CD-adapco

- **Exploration:**
  - Tile tilt angle change from 1 deg to 16 deg shows reduction in both CO and NOx .
  - Reduction in overall flame height and volume seen with increasing tile angle
  - Fuel injector spacing affects flame height and flame volume. This in turn affects the emission values
- **Optimization**
  - Best design that results is minimum CO and NOx values is Design 32.
  - CO reduced by 4% from baseline
  - Simultaneous NO reduction by 11% from baseline

401341 Modeling, Optimisation and Control of Solar Absorption Air-Conditioning System

Abstract:

Solar thermal energy can provide a benign and environmentally friendly heat source for driving a vapour absorption refrigeration system of unit capacity using R 717 (NH3) and water as the working fluids. The last few decades witness an upsurge in energy demand, both at the domestic and industrial cooling and heating demand. Traditionally, heating, ventilation and air-conditioning (HVAC) performance and design are based on construction of prototypes in a wind tunnel. This approach is not only time consuming but expensive. In this study, a model-based approach for the modelling, optimisation and control of a solar vapour absorption air-conditioning system is presented. A model-in-loop and hardware-in-loop model was also develop to validate the control algorithm.

401412 New Development on Hoist Scheduling for Multi-Stage Material Handling

Abstract:

New Development on Hoist Scheduling for Multi-stage Material Handling

HonglinQu, Chuanyu Zhao and QiangXu*

Dan F. Smith Department of Chemical Engineering

LamarUniversity, Beaumont, TX77710, USA

Abstract

Multi-stage material handling (MSMH) processes are widely used in industries for the manufacture of various workpieces and products (jobs) in large scale. Generally, a hoist, as control device, will handle all jobs by following a preset movement schedule, including joblifting, releasing and moving along a production line. Lots of engineers have studied the feasibility and optimization of hoist movement schedules. However, the productivity of an MSMH process not only depends on the hoist schedule, but also relies on the production line arrangement, such as the spatial allocation of processing units. During production, hoist will travel forth and back to pick up job from its current unit and release it into the next one, following its processing requirement. In this way, a proper arrangement of unit position would dramatically shorten the total moving time and improve the processing capability. As far as we know, no work has been done in this field. Therefore, we try to arrange the unit allocation in a more compact way to optimize the hoist schedule.

In this paper, a two-dimensional (2D) matrix arrangement of processing unit has been proposed to replace original one-dimensional (1D) line arrangement for the optimization of productivity, and a mixed integer linear programming (MILP) model has been developed to solve this problem. The major advantage of this study is that the compacted allocation provides shorter paths for hoist to move, so the productivity could be improved significantly. The efficacy of the proposed methodology is demonstrated by a case study. The comparisons between 1D and 2D production line arrangements have also been conducted.

401413 Simultaneous Scheduling of Front End Crude Transfer and Refinery Processing (Electronic Poster Presentation)

Abstract:

Simultaneous Scheduling of Front End Crude Transfer and Refinery Processing

Jialin Xu, Shujing Zhang, Jian Zhang, and Qiang Xu

Dan F. Smith Department of Chemical Engineering

Lamar University, Beaumont, TX 77710

Abstract

Scheduling and planning of crude oil is very important to a petroleum refinery due to the potential realization of large cost savings. It is also the one of most challenging problems for refinery operations. In this paper, a new continuous crude scheduling model has been proposed to simulate and optimize the front-end and refinery crude-oil operations. This scheduling problem involves crude oil unloading from vessels to the crude storage tanks at onshore berths, transfer of crude from these tanks to the charging tanks, charging crude distillation units with crude mixes from the charging tanks and further processing of crude inside the refinery, which includes crude distillation, cracking, coking, reforming, hydrotreating, product blending and component recovery.

The objective of the proposed crude scheduling problem is to maximize the total operational profit; meanwhile, it should meet inventory and production demands, and try to maintain refinery plant's normal operations under uncertainties. The scheduling model is a large-scale mixed integer nonlinear programming problem(MINLP). An outer-approximation based decomposition method is implemented to obtain the optimum solution of the developed MINLP scheduling model.

401527 An Optimization Framework for the Day-Ahead Scheduling of District Cooling Systems

Abstract:

The main goal of most district cooling systems is to centrally produce and distribute chilled water at a minimum cost while satisfying demand. Chilled water is produced by production plants and distributed to buildings equipped with energy transfer stations (sub-stations). After the heat transfer takes place, water is returned to the production plants where the chilled water is regenerated. District cooling is essential for sustainable economic and urban development. The reason for this is that by centralizing the chilled water generation, district cooling systems can increase system efficiency, reduce air pollution, decrease emissions of refrigerants, and manage electricity demand on-site and to the electric grid as a whole.

Chillers are the main components of the chilled water production plants. The production cost of chilled water plants that use electrically powered chillers is deeply sensitive to electricity price variability [4]. To minimize the production cost, thermal energy storages are often used [1,3]. The main idea is to store chilled water in the thermal energy storage when the electricity price is low and discharge it when the electricity price is high. Although this operating rule may seem trivial to implement, multi-modal electricity price curves, thermal energy storages with limited capacity as well as multi-period constraints on chillers present a challenge to the operators who aim at developing the day-ahead operating schedule of the cooling system at a minimum cost.

In this work we propose an optimization framework that deals with this situation. We show that by combining the latest results on piece-wise linearization techniques [2] with disjunctive programming strategies [5] we are able to generate an optimization model which can be solved fast and efficiently as a mixed-integer problem. We describe how this framework is implemented and successfully used to calculate the day-ahead optimal schedule of the thermal energy storage and chillers at the Thermal Energy Corporation, a district energy company located in Houston, TX.

[1] W. Bahnfleth and W. Joyce. 1994. Energy use in a district cooling system with stratified chilled water storage. ASHRAE Transactions. Volume 100, pp 1767-1778

[2] B. Geißler, A. Martin, A. Morsi, L. Schewe. 2012. Using Piecewise Linear Functions for Solving MINLPs. The IMA Volumes in Mathematics and its Applications. Volume 154, pp 287-314

[3] W. J. Cole, K. M. Powell and T. F. Edgar. 2012. Optimization and advanced control of thermal energy storage systems. Chemical Engineering Reviews. Volume 28, pp 81–99

[4] K. M. Powell, W. J. Cole, U. F. Ekarika, T. F. Edgar. 2013. Optimal chiller loading in a district cooling system with thermal energy storage. Energy. Volume 50, pp 445–453

[5] J. P. Ruiz, J. Wang and C. Liu. 2013. Logic-based outer-approximation method for the security constrained unit commitment problem. IET Generation, Transmission & Distribution. Volume 7, pp 1210-1218

401529 Optimized Control Strategy of Distillation Column Under Startup and Shutdown

Abstract:

The startup and shutdown operations of a distillation column involve complex heat and mass transfer phenomena in different transient time periods. Dynamic simulation can be used to study and analyze these dynamic behaviors. Based on dynamic simulations, column operating procedures and control strategies can be improved to guarantee the operations flexible and reliable. And operation time and energy savings have an appreciable economic significance.

In this study, rigorous dynamic simulations are employed to investigate the startup and shutdown of C2 splitter in an ethylene plant. First, in the base case, the dynamic model for the column is built with normal control strategy. All the operating procedures are programmed and automated in one rigorous simulation model. Based on the results in the base case, the control strategy is improved in the optimized case. Second, dynamic model for C2 splitter is built with optimized control strategy. Split range control and low select control are applied in this optimized control strategy. The automation for start-up and shutdown operation is enhanced. In summary, dynamic simulations performed in this study will help identify operating procedures and control strategies to start up or shut down a distillation column quickly and safely, so as to save plant raw material and energy, reduce operational risks, and thus enhance the industrial sustainability.

401556 On-Line Optimal Control of Molecular Weight Distribution in Batch and Semi-Batch Free Radical Polymerization Processes

Abstract:

Polymer molar mass distribution is one of the most important polymerization control variables that affect many polymers' characteristics such as strength and thermal stability. Although there have been full feedback control of reactors on small molecules, polymerization reaction control is less advanced due to lack of understanding of the dynamics of the process, inadequate in situ sensors and nonlinearity in the behavior of the reaction for instance gel affect that influences kinetic parameters during the reactions.

This contribution presents a complete real time feedback framework to control the conversion kinetics and polymer chain length distribution during the free radical batch polymerization operation. In the first step the process is conducted in batch mode where a detailed polymerization model is applied along with the method of finite molecular weight moments to determine a sequence of reactor temperature set points which lead to the desired molecular weight distribution. The optimization of the model is performed using the gOPT function in gPROMS. In order to have a better controllability on the process the system is also simulated in semi-batch mode where reagents are fed into the reactor during the reaction. The formalism in this case combines kinetics with time dependent processes related to flows of material into and out of the reactor. The objective is to find the monomer (initiator) flow rate to control the average molecular weight and the temperature is adjusted to manipulate the polydispersity. To validate the model and assess the merits of the proposed strategy, free radical polymerization of methyl methacrylate is carried out using 2,2'-azobis(2-methylbutanenitrile) as initiator. Automatic Continuous Online Monitoring of Polymerization reactions (ACOMP) is used to follow the conversion and evolution of the average mass distribution, and multi-detector SEC is applied to cross-check results and measure full distributions of end products.

401655 DNA Hybridization on Silica Surface for Gene Chip Design Studied with Molecular Dynamics Simulations

Abstract:

Nucleic acid hybridization behavior has attracted research attention due to their unique pairing interactions. The short nucleic acid probes such as miRNA or 16s regions of microbial genomes are being applied as biomarker for the detection of various diseases, such as cardiovascular or infectious diseases. To facilitate our experimental design of gene chip for early diagnostic purpose, we performed a fully atomistic Molecular Dynamics simulation to study DNA hybridization behavior on α-cristobalite (101) surfaces, which is partially hydroxylated and also chemically modified with aldehyde chains. Drawn on the atomistic simulations, we investigated the morphology of the grafted DNA strands on the surface and the interactions between DNA complementary strands, as a function of the surface properties and solution environment (such as ion strength) for the purpose of enhancing hybridization efficiency. The current research will also lay the foundation for future research with the coarse-grained modeling.

401676 Structures, Dynamics, and Water Permeation Free Energy Across Bilayers of Lipid a and Its Analog Studied with Molecular Dynamics Simulation

Abstract:

Fundamental studies of the supramolecular layer structures, dynamics and water permeation free energy of hexa-acyl-chain Lipid A and its analogue of tetra-acyl chains would be useful for polymer membranes design for endotoxin removal in water treatment, drug delivery and other biotechnologies. In this work, we studied their supramolecular bilayer by using molecular dynamics simulations and efficient free energy computations. Our simulation accuracy was verified by the agreement between the bilayer structural properties (structure factor, bilayer thickness, and the area per lipid) and lateral diffusion coefficient in our simulation and experimental measurements. More importantly, our simulation for the first time illustrated hexagonal compact packing of the hydrocarbon acyl chains within a leaflet of Lipid A membrane (at 298 K and water content of 40 wt %), which is consistent with experiments. In contrast, Lipid A analogue is found with less ordered ripple structures at the same condition. Our study also demonstrated slower dynamics and larger and broader free energy barrier (23 kJ/mol) for water permeation for Lipid A, compared with that of Lipid A analogue. Moreover, the analysis of dynamics showed that highly hydrated hydrophilic diglucosamine backbone is structurally stable, whereas the interdigitated hydrophobic acyl chain tails inside the membrane with faster dynamics screen the aqueous environment from the lipid interior and also reinforce the membrane's structural stability.

401681 Lysozyme Adsorption and Desorption on the Azobenzene Self-Assembling Surfaces Studied with Molecular Dynamics Simulations

Abstract:

The reversible photosponsive isomerization of the azobenzene and its derivatives can control protein structure in an aqueous environment with the alternation of visible and UV lights for very promising applications in drug delivery and biomaterials. However, protein adsorption/desorption behavior on the azobenzene self-assembled monolayer surface (Azo-SAMs) is still unknown. In this study, atomistic molecular dynamics simulations were performed to study lysozyme adsorption and desorption on a silica surface grafted with Azo-SAMs. Protein structural arrangements (i.e., the shape and secondary structures), its mobility and the interactions with the surfaces, as a function of the azobenzene SAMs density and morphology, are systematically investigated.

401708 Improving Process Sustainability and Profitability for a Large U.S. Gray Iron Foundry (Electronic Poster Presentation)

Abstract:

Improving Process Sustainability and Profitability for a Large U.S. Gray Iron Foundry

Prashant Nagapurkar, Shyam Paudel and Joseph D. Smith, PhD.

Missouri University of Science and Technology, Rolla, Missouri, USA

Abstract

Energy savings and sustainability has been an important topic in many industrial processes. Limited energy resources and increasing electricity prices make foundry processes less competitive in today's world economy. For foundry processes, the energy costs generally represent 5-7% of the total sales. Most of the energy used in an foundry is directly related to melting and heat treatment operations. Integrating heat and power systems in a foundry through waste heat recovery techniques can lead to energy reutilization that improves process sustainability and profitability. The work presented here is an in-depth energy management analysis using Powersim^{TM} to identify potential energy savings through process optimization and advanced control strategies. A system dynamics model based on the interactions among the major energy intensive variables has been developed using Powersim^{TM} to estimate energy consumption. Our dynamic model estimates the monthly energy usage and its costs for a full calender year. The estimated energy consumption and its costs have been validated by comparison to the actual plant data obtained from the Mapleton foundry.

The average metal caster has a 2.4% pre-tax operating profit on sales. This profit is susceptible to fluctuating utility costs and process disturbances. Since foundry operates with very thin profit margins it is imperative to ensure that the foundry is running with efficient process control. Hence, advanced control strategies for the melters, dust collectors, fans and burners have been proposed and implemented in the Powersim^{TM} model for optimum energy usage. Our results show that these control strategies along with waste heat recovery techniques could save nearly 18% in energy savings and $1.8 million in costs annually for the Mapleton foundry.

400426 Synthesis of Efficient Multicomponent Separation Processes

Abstract:

Separation processes find application in a variety of industries like the chemical, petrochemical, pharmaceutical, gas separation and semiconductor industries. It is estimated that separation processes account for nearly 40-70% of chemical plant costs. One reason for the associated high costs is that the industry relies on the experience and intuition of a few expert engineers to build separation sequences for any given application. In this work, we present a systematic method to design a separation system that is up to 50% cost-effective and energy-efficient over existing systems. Our focus is on distillation configuration design. First, we enumerate all useful distillation sequences that are possible for separating a given mixture, and then incorporate them into an optimization framework which produces a rank-list of distillation sequences based on a desired criterion (energy consumption/overall cost/thermodynamic efficiency). Such a tool in the hands of an industrial practitioner will allow the person to focus on a certified small subset of useful configurations. Furthermore, we present some exciting recent progress made in process intensification for distillation, in areas of dividing wall columns and simultaneous heat and mass integration.

398087 On Diffusion Coefficients, Relaxation Times of Oligonucleotides

Abstract:

DNA microarrays have been applied to gene expression, genotyping and sequencing. The hybridization step between probe and target molecules may take several hours. The understanding of the hybridization kinetics is important. Some investigators have found the hybridization overall reaction rate to be reaction rate limited in the 90s and some other investigators in the early 00s have found the overall hybridization reaction rate to be diffusion limited (Pascault and Zhou, 2008). They go further to improve the overall hybridization reaction rate by providing convective mass transfer to decrease the diffusion limitations. They obtain these recommendations after performing computer simulations by finite element numerical solutions using Comsol Multiphysics software package. The length of the DNA target molecules range between 50 – 350 bp. The probe molecules are shorter than the target molecules and have about 10-60 bp of Adenine, Cytosine, Guanine and Thymine. The surface concentration of probe molecules range between 1.2 x 10 $^{-10}$ – 4 x 10^{13} probes/cm^2 or 1 attomoles/m^2 to 0.1 micromoles/m^2. The diffusion coefficient values for the base pairs decreased with the length of the oligonucleotide. The variation of diffusivity values as a function of the length of the oligonucleotide from different investigators is shown in Figure 1.0. With a R^2 value of 0.9956 the following curve could be seen from the trendline except for the data point from Ferrari and Bloomfied given in Pascault et. Al. [2008]. Their value is also given as a red rectangle in Figure 4.3. The diffusion coefficient can be written as follows;

$$D_{pl} = al^{0.639}$$

The relaxation time for oligonucleotides can be written as;

$$\tau_r = b(l^{1.639})/RT$$

Where l is the lengt of the oligonucleotide in terms of number of base pairs, and D_{pt} is the binary diffusivity of the probe molecules in target molecules. A_1 was found to be 6 x 10^{-10} m^2/s from the regression tool in MS Excel 2010 for Windows using a HP Compaq Elite 8300 desktop computer with i7 intel processor. R the universal molar gas constant, T the absolute temperature. Target molecules of 1 – 1000 pM were considered. The channel height of the microfluidic elements range was 50 – 100s μm. Target microarray spot size considered was 200 μm. The two dimensional microfluidic chamber had a height of 100 μm.

Figure 1.0 Diffusion Coefficient of Oligonucleotide as a Function of Size

399668 Thermodynamic Modelling of Br2-HBr-H2O Ternary System with Enrtl Model

Abstract:

The storage of energy has always been researchers' concern for decades. As an improvement of fuel cells, the technique of redox flow cells provides us a great balance between energy density and electric-to-electric efficiency. The H_2-Br_2 flow cell system has the advantages in reaction kinetics, power density, energy storage capacity, simplicity, and low cost on materials. In the system, there are molecular components including H_2, Br_2, HBr, H_2O, and ionic components such as H_3O^+, Br^-, even H_5O_2+. Proper thermodynamic modelling of the complex electrolyte solution is of great importance for further development of the H_2-Br_2 battery system. Wozny looked into phase equilibria of HBr/H_2O system, and a correlation for only pressure was put forward in 1981. Applying electrolyte non-random two liquid model, this work investigates the adjustable pair parameters between molecular-molecular components, molecular-ionic components, and ionic-ionic components. Experimental data of phase equilibria, solubility, and molar enthalpy are used to regress these parameters for both HBr-H_2O binary system and Br_2-HBr-H_2O ternary system. It turns out that ENRTL model works in excellent precision for this system. The redox potential of flows in the cell system can also be evaluated with this model.

400429 Evaluation of the Methods for Predicting Liquid Densities of High-Temperatures and High-Pressures Deep Petroleum Formations

Abstract:

Six currently used methods for predicting liquid densities of high-temperature and high-pressure (HTHP) deep-water petroleum formations are evaluated for ease of implementation and applications for flow assurance modeling, pipeline flow simulation and reservoir simulation of carbon dioxide sequestration in the depleted oil-reservoirs. The evaluated techniques include the various modified versions of the *Peng-Robinson EoS*, the *high-temperature, high-pressure, volume-translated Peng-Robinson EoS*, the *volume-translated Soave-Redlich-Kwong EoS*, the *generic Van der Waals cubic EoS*, the *modified Tait equation* and the *perturbed-chain statistical associating fluid theory EoS*. The HTHP liquid density data used in the evaluation of the methods for HTHP liquid densities are reported in the literature by Enick Group at U of Pitt, Caudwell-Trusler-Wakeham Group at Imperial College London, Ducoulombier et al. in France, the Doolittle specific volumes of n-Alkanes (nC_7 to $C_{40}H_{82}$) and the Sage-Lacey volumetric data of CO_2, N_2, H_2S, C_1 to nC_{10}, including the isomers of iC_4 and iC5. While all the evaluated methods gave various degrees of uncertainties at HTHP conditions (due to the judicious choice of the parameters required to implement PC-SAFT and the volume-translated techniques), the VDW generic equation of state (with covolume temperature function) performs better and in overall agreement with the measured liquid densities at the HTHP conditions of deep-water petroleum formations.

The analysis of model results highlight the capability of the various techniques to account for the liquid densities measured for some isomers investigated at HTHP conditions as well as for high-molecular-weight fluids. The HTHP VT-PREoS provides minor improvements over the original 1976 design of PREoS, but neither EoS is capable of accounting for the effect of isomer structural differences on the measured liquid densities at HTHP conditions and significant deviations are observed at HTHP liquid densities of heavier molecular-weight hydrocarbons. However, the PC-SAFT EoS, with pure component parameters derived from the literature provides modest improved liquid density predictions over those obtained with the PREoS or HTHP VT-PREoS or volume-translated SRK EoS.

Overall, the VDW generic cubic EoS (with embedded covolume temperature function) provides more accurate predictions of liquid densities at HTHP conditions associated with the deep-water petroleum formations than all the methods of predicting liquid densities that are analyzed in this investigation. Even though the PC-SAFT EoS shows improved predictions of liquid densities over the Tait correlations and over those EoS (without the embedded covolume temperature function), it still over predicts the experimental liquid densities at HTHP conditions. The VDW generic cubic EoS is recommended for all analyses of flow assurance modeling. But, we are still performing similar analysis for derivative properties (isothermal compressibility, volume expansivity, bulk modulus, isobaric and isochoric heat capacities, Joule-Thomson coefficient and speed of sound) at HTHP conditions associated with the deep-water petroleum formations.

Keywords: cubic equation, Tait equation, liquid density, covolume temperature, heavy hydrocarbon

Thermodynamic Model for Quaternary Aqueous Ca^{2+}, Na^{+}, K^{+}, Cl^{-} System

Abstract:

Symmetric Electrolyte Non-Random Two Liquid model (eNRTL) is selected as the thermodynamic framework to model the quaternary aqueous Ca^{2+}, Na^{+}, K^{+}, Cl^{-} electrolyte system. This is a part of a large effort to develop a comprehensive thermodynamic model for high salinity produced water in oil and gas production. Each binary system (electrolyte-water or electrolyte-electrolyte) needs three pairs of binary parameters which are directly correlated with excess Gibbs energy, excess enthalpy and excess heat capacity of the system concerned. These binary parameters are regressed with thermodynamic and phase equilibrium data including salt solubility data available in literature. Binary parameters are obtained for $CaCl_2$-water, $CaCl_2$-KCl and $CaCl_2$-NaCl binary systems. These binary parameters are integrated with the rest of the parameters readily available in literature. The model shows excellent agreement with literature data with temperature from 0°C - 200°C and electrolyte concentration up to saturation. This model is being extended to cover Mg^{2+}, Ba^{2+}, Sr^{2+}, SO^{4-} and $HCO3^{-}$.

399251 Understanding the Effect of Inhibitors on Controlling Asphaltene Precipitation and Aggregation Phenomena

Abstract:

Understanding the Effect of Inhibitors on Controlling Asphaltene Precipitation and Aggregation Phenomena

Mohammad Tavakkoli[1, 2], Miguel Garcia-Bermudes[1], Ariana Melendez[1], Caitlin Garcia[1], Sarah Correa[1], Quinton Cox[1] and Francisco Vargas[1, *]

1. Department of Chemical and Biomolecular Engineering, Rice University, Houston, Texas, USA.

2. Department of Chemical and Petroleum Engineering, Sharif University of Technology, Tehran, Iran.

* fvargas@rice.edu

Asphaltenes tend to precipitate and deposit during oil production and clog the wellbore and production facilities. This can cause an excessive downtime and high cleaning costs running into several millions of dollars. Therefore, a mitigation technique such as using asphaltene inhibitors is necessary. To improve the efficiency of the current inhibitors or develop a new generation of asphaltene inhibitors, it is necessary to study the effect of inhibitors on controlling asphaltene precipitation, aggregation and deposition processes.

In this study, we have investigated the effect of different commercial inhibitors on asphaltene precipitation and aggregation processes. It needs to be clear whether the inhibitors make changes on the asphaltene precipitation such as changes on the onset and the amount of precipitation or they only affect the asphaltene aggregation process by dispersing asphaltene aggregates and changing the particle size distribution. To do so, three commercial inhibitors were used and asphaltene precipitation and aggregation phenomena were studied using two different techniques: "Direct spectroscopy" and "indirect method". The experiments were conducted at ambient conditions and asphaltenes were precipitated using light normal alkanes such as n-pentane and n-heptane. Using direct spectroscopy technique, we found that asphaltene particle size distribution changes by adding the inhibitors to the crude oil especially at high amount of the added precipitant. The inhibitors break the asphaltene agglomerates and form smaller size aggregates. Indirect method, which is a combination of gravimetric and spectroscopy methods and is a novel technique for studying asphaltene precipitation, showed that there are no changes on the onset and the amount of asphaltene precipitation when the inhibitor is added to the crude oil prior to adding the precipitant. Therefore, one can conclude that the commercial inhibitors can disperse the already precipitated and aggregated asphaltenes and form smaller particles, but it is still under investigation whether this working mechanism of inhibitors always prevents asphaltene deposition in the wellbore or may worsen the problem in some cases.

Keywords: Asphaltene Precipitation, Asphaltene Aggregation, Commercial Inhibitors, Indirect Method.

400030 PVT Modeling of Crude Oils: Development of Advanced Characterization Methods and Accurate Prediction of the Onset of Asphaltene Precipitation

Abstract:

With ever-increasing complexities in production environments – including gas injection for EOR or deep sea Pressure-Temperature conditions – and the high cost of sampling reservoir fluids, accurate modeling of PVT properties with minimal field data is critical. These difficulties necessitate a continual reevaluation of the characterization methods for crude oils as well as the Equation of State (EoS) models used to predict PVT properties. The impact of such accurate PVT characterization and modeling of crude oils in reservoir simulations and EOR applications cannot be over-emphasized. In this work, we review well known methods for characterizing the plus fraction of black oil and gas condensate mixtures and compare various equations of state in predicting the phase behavior of the reservoir fluids in a wide range of pressures and temperatures. The equations of state studied in this work are the conventional cubics (i.e. Soave-Redlich-Kwong and Peng-Robinson) with and without volume correction and PC-SAFT. We will revisit previous characterization methods and discuss the development of new, simplified characterization methods. A thorough discussion is presented on the characterization procedures, advantages and disadvantages of the different methods, and the performance in the prediction of PVT properties.

From the flow assurance stand point, accurate prediction of asphaltene precipitation is absolutely critical for (a) understanding the likelihood of asphaltene deposition problems, (b) implementing proper mitigation strategies and (c) developing suitable EOR strategies, as gas injection can potentially increase the propensity of asphaltene precipitation. In this regard, we build on our previous work in the use of PC-SAFT EoS along with simplified correlations relating the PC-SAFT parameters of Asphaltenes, Aromatics and Resins to their MW (or a MW distribution) and a new parameter called aromaticity. Further, the accuracy of newly developed characterization methods in the prediction of bubble pressure curves and asphaltene onset curves will be discussed.

401699 Hydrate Plug Location during Depressurization Remediation Technique

Abstract:

During the remediation of hydrate blockage in a pipeline, certain key parameters such as the location of the first plug, the velocity of the plug upon release, the gas volume fraction upstream and downstream of the pig, etc. are important to know for safe handling of the process. Typical remediation of a blocked pipeline would include alternating cycles of depressurization and inhibitor injection to safely handle plug movement.

Utilizing operating information during the remediation process, MSi Kenny has developed a toolkit that estimates the hydrate plug location based on alternating between inhibitor injection and depressurization method of remediation. The purpose of this model is to determine the plug location and calculated downstream gas volume fraction.

The model uses the pressure rise in the pipeline during the inhibitor injection cycle (with known methanol injected volume), the subsequent pressure drop and liquid offtake volume during the depressurization step. Using these inputs the model iteratively calculates the first mathematically stable plug location and downstream gas volume fraction. These calculations are currently based on basic equations of state. Initial comparison of the model to multiphase flow simulations show the results to be in close agreement.

This model offers a tool to predict potential hydrate plug locations for this particular remediation technique. With an estimate of the hydrate plug location, operators can have a better idea of hydrate plug movement risks, more specific hydrate remediation locations, or simply a secondary method to verify plug locations.

400322 Microfluidic Systems That Mimic Reservoir Conditions for Visualizing Asphaltene Aggregation and Deposition Near Wellbore Conditions (Electronic Poster Presentation)

Abstract:

Asphaltene deposition is well known as a cause of production rate decline and various other operational problems in most oil production processes. Asphaltenes are typically stable under virgin reservoir condition, however may be destabilized and precipitate from crude oil and deposit on the internal surface of equipment or near-wellbore reservoir during production and transportation. Asphaltenes precipitation and deposition are often due to changes in thermodynamic properties, and pressure has a large effect on the flocculation of asphaltenes. Previous studies have investigated these phenomena in bulk volume (e.g. in stainless steel pipe) but visualization are often missed in these "black-box" type experiments, and the results from such macroscopic observations could not reflect the nature of asphaltenes at micro-scale. Micro-scale behaviors of asphaltenes are important for the fundamental understanding of origins of asphaltene flow assurance problems. Prevention and mitigation of asphaltene problems may also need to be applied at the source of crude oil in wellbore conditions and in porous media. Hence there is a great need to fill in the gap by visualizing asphaltene aggregation and deposition behaviors at micro-scale near wellbore conditions. In this report, we will demonstrate preliminary results of such experiment using microfluidic systems. The variation of asphaltene solubility in model oil with change of pressure will be visualized in micromodel, and flow experiment of asphaltene deposition on porous media surface will also be demonstrated. These microfluidic experiments has the potential to provide new information for fundamental research at wellbore conditions as well as a mean of fast inhibitor development.

400338 Flowloop Measurements of Hydrate Blockage Risk in â€"Under-Inhibited Gas-Dominant Systems

Abstract:

In conventional flowlines, gas hydrate formation is mitigated through the injection of thermodynamic hydrate inhibitors (THIs) that shift the hydrate stability criteria toward high pressure and lower temperature. Complete hydrate avoidance with THIs substantially increases infrastructure and operating costs, which may affect the economic viability of deepwater assets. Reducing the rate of THI injection below the complete inhibition threshold may translate to significant cost savings, but adoption of this practice requires a rigorous experimental confirmation that such under-inhibition will not increase the risk of hydrate blockage formation. In this study, we have deployed a high-pressure sapphire autoclave and high-pressure single-pass flowloop to measure hydrate formation and resistance-to-flow (motor torque or pressure drop) in a matrix of under-inhibited conditions. 20 autoclave studies were performed with deionised water containing 0-20 wt% monoethylene glycol (MEG). The rate of hydrate growth and severity of blockage formation decreased monotonically with the initial MEG fraction; visual observation of the sapphire cell suggested that MEG reduced hydrate film growth and particle deposition on the wall. The hydrate growth rate was accurately predicted within 2 vol% by a kinetic growth model. 20 gas-dominant flowloop experiments were performed with 10 vol% water flowing under annular conditions. Increasing the amount of MEG in water monotonically decreased (i) the hydrate growth rate, (ii) the rate of pressure drop increase, and (iii) the frequency of sloughing events. At subcoolings below 5 K, the pressure drop increase due to wall deposition and the global hydrate growth rate were compared. The results suggest that approximately half the hydrate was deposited on the pipeline wall; further studies are underway to expand this understanding to systems at higher subcooling. Together, these experimental studies demonstrate that a reduction in MEG fraction of up to 10 wt% from the complete inhibition requirement did not result in an imminent increase to hydrate blockage risk. The hydrate growth rate may be readily estimated through a kinetic model, where MEG fractions above 10 wt% were consistently observed to improve hydrate transportability.

400350 A New Quantitative Approach to Assess the Risk of Hydrate Blockage Formation in Subsea Flowlines

Abstract:

As production systems move toward deeper water, high operating pressures and longer thermal exposure times will significantly increase the risk of gas hydrate formation and potential flowline blockage. Over the past two decades, thermal and chemical hydrate management strategies have been developed with a goal of decreasing capital and operating costs while minimising the risk of hydrate blockage formation. The successful adoption and refinement of these management technologies hinges on accurate predictive capabilities for hydrate formation and transportability. Current simulations of hydrate blockage formation rely on a set of deterministic calculations, which do not account for distributed phenomena such as hydrate nucleation. In this presentation, we introduce a new risk engine that is designed to couple probabilistic phenomena – such as hydrate nucleation, deposit sloughing, and jamming – with existing deterministic models. The current incarnation of this engine utilises the Hydrate Flow Assurance Simulation Tool (HyFAST) to describe deterministic phenomena, where the severity of hydrate blockage formation is quantified through the increase to hydrate slurry viscosity for water- and oil-continuous systems. The first deployment of this risk engine incorporates a new probability distribution to describe hydrate nucleation as a function of system temperature; thousands of hydrate nucleation events were observed in an Automated Lag Time Apparatus (ALTA) to inform this distribution. The results of this pairing demonstrate a new ability to quantify the risk of observing hydrate blockage formation in a subsea flowline. In this presentation, we will apply this risk engine to a hypothetical long-distance tieback. The results demonstrate how flow assurance engineers can decide upon an acceptable risk tolerance (e.g. 90% of observations) and deploy this simple calculation strategy to optimise subsea design decisions, such as insulation thickness, flowline diameter, and choke opening.

400359 Prediction of Asphaltenes Solubility in Organic Solvents Via NRTL-SAC and COSMO-SAC

Abstract:

Asphaltenes are polyaromatics with hetero atoms (N, O, and S) that contribute to the polar characteristics of hydrocarbon mixture, i.e., crude oil. Solubility of asphaltenes reflecting the characteristic nature of mixture of compounds collectively, i.e., hydrophobicity, polarity, and hydrophilicity, greatly affects the production and transportation of hydrocarbon mixture, and causes fouling in unit operations. Current solubility models are based on the Scatchard–Hildebrand concept, and further extended with the Flory–Huggins correlation. These model require molecular weight and molar volume as input parameters. However, asphaltenes are solubility class rather than pure components, and the uncertainty of distribution of pure components within asphaltenes hampers the efficacy of these models. In the present work, 'like dissolves like' phenomena and solvation thermodynamics are used to calculate the activity coefficient of asphaltenes and solubility subsequently. Also, the conceptual segments of NRTL-SAC model for asphaltenes from published solubility data in mixed solvents is identified to generate the apparent sigma profile, i.e., a histogram of charge density distribution over the molecular surface of asphaltene. The sigma profile is then used in COSMO-SAC model to calculate activity coefficient and solubiliy. By comparison, the sigma profiles of asphaltenes is also obtained from quantum chemistry calculation, i.e., density functional theory, with molecular structure for asphaltene molecules as input collected from literature proposed by different authors. Furthermore, various entropy models are investigated since both Flory-Huggins and Staverman-Guggenheim models have been found with issues.

400387 Design for Mechanical Integrity in Thermal Mixing Points

Abstract:

Design for mechanical integrity in thermal mixing points

Authors: *Santhosh K Shankar*[*], *Gonghyun Jung*[^], *Karl Anderson*[*], *Greg Bethke*[*]

Thermal fatigue can occur near mixing points where fluids of different temperatures comingle to reach a desired temperature for downstream processing requirements or in hydro-processing units where hydrogen is injected to treat a hydrocarbon stream at a different temperature. Failures due to thermal fatigue may lead to loss of primary containment, resulting in fire, health hazards and production deferment resulting from unplanned unit shut-downs and are hence associated with problems of high business impact. Thermal sleeves, quills and/or spray nozzles may be used in mix points to protect the walls of the piping against fatigue. The primary challenge towards design of thermal sleeve is- understanding the process of thermal mixing near and downstream of the mixing point under different flow regimes. Identification of the appropriate dimensional and non-dimensional quantities on which the thermal mixing process depends is hence critical. The present work showcases an attempt to identify key universal parameters that govern the thermal mixing process.

Towards this objective, a parametric study was conducted using computational fluid dynamic (CFD) simulation at varying process conditions to understand the thermal behavior of the flow. Temperature statistics are probed along the wall at downstream and azimuthal locations of the pipe to map trends in the thermal behavior of the flow.

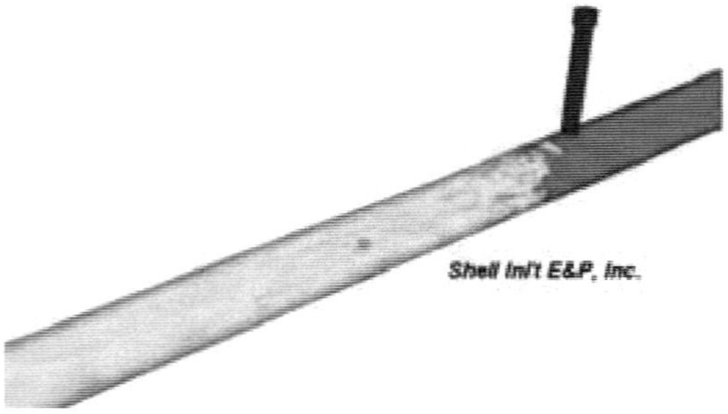

Figure: Temperature fluctuations along the pipe walls encountered in a mixing point from mixing of hot (red) and cold (blue) fluids.

399581 Stimuli-Responsive Rheoreversible Fracturing Fluids for Unconventional Oil and Gas Extraction

Abstract:

Even the most pessimistic US Energy Information Administration (EIA) models predict a near-doubling of U.S. production of tight oil by 2035 as crude production from other sources remains static or diminishes. As a result, the development of specialized fracturing fluid systems has been a constant research and development focus driven by the strong need for optimization of fracture propagation and permeability enhancement. To optimize and lessen environmental impact of reservoir stimulation, an environmentally benign, CO2-activated, rheoreversible fracturing fluid is presented. This fracturing fluid enhances permeability due to in situ volume expansion triggered by the presence of CO2. Our results show **100 % volume expansion** of PNNL fracturing fluids at temperatures relevant to tight oil reservoirs. This rapid volume expansion created fracture networks in highly impermeable crystalline rock cores with **permeability values up to 3 orders of magnitude higher** than the permeabilities obtained when using conventional fracturing fluids. The chemical mechanism, phase-change behavior, and rheology of this fracturing fluid at tight oil temperatures and pressures are presented.

Mitigating Downhole Flow Assurance Issues Using Specialist Modeling Software

Fabin Xu[1], Chunmin Zeng[1], Chong Zhang[1], Zhao Dong[1], Xiaohong Zhang[2], Adrian Ferramosca[3], Behnam Salimi[2] and Richard Szczepanski[2]

(1) CNOOC Zhanjiang Exploration and Development Research Institute, Zhanjiang, China

(2) Infochem PVT Physical Property and Thermodynamics, KBC Advanced Technologies plc, London, United Kingdom

(3) Weatherford, Edinburgh, United Kingdom

We are currently exploiting a large oilfield in the Beibu Gulf, South China Sea, requiring the design of 12 wells. We are using specialized well-modeling software using advanced flow assurance techniques incorporating accurate wax, asphaltene and hydrate prediction methods to revise our previous design to include additional solids deposition mitigation techniques, including thermal insulation in a number of wells where this was not originally thought to be a problem.

Fluid Description

Source Composition

Fluid Component	MolarComposition (Fraction)
H2O	0.0098
N2	0.0108
H2S	0.0000
CO2	0.0302
C1	0.5103
C2	0.0244
C3	0.0616
IC4	0.0059
NC4	0.0124
IC5	0.0483
NC5	0.0000
C6_1	0.2442
C6_2	0.0419

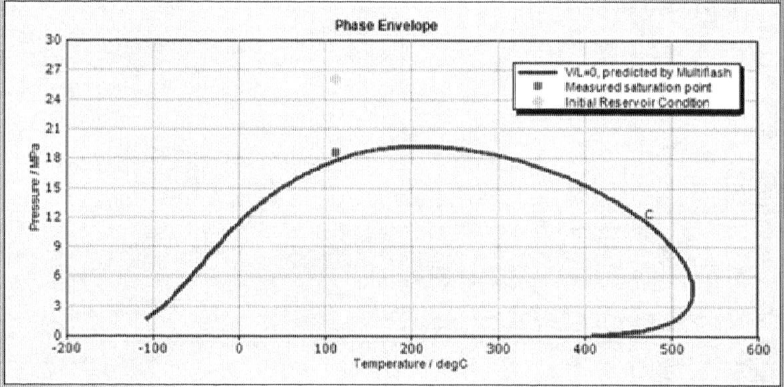

Software used

WellFlo™ v6.0 (Weatherford)

PVTFlex™ v2.0 (Weatherford)

Multiflash™ v4.4 (KBC-Infochem)

Weatherford's WellFlo™ well design and evaluation software, through its direct link to KBC-Infochem's Multiflash™ rigorous wax, hydrate and asphaltene flow assurance models, is able to simultaneously plot the solids formation phase envelope together with the pressure-temperature profile followed by the hydrocarbon fluids from bottom-hole to wellhead to platform.

WellFlo™ Plots

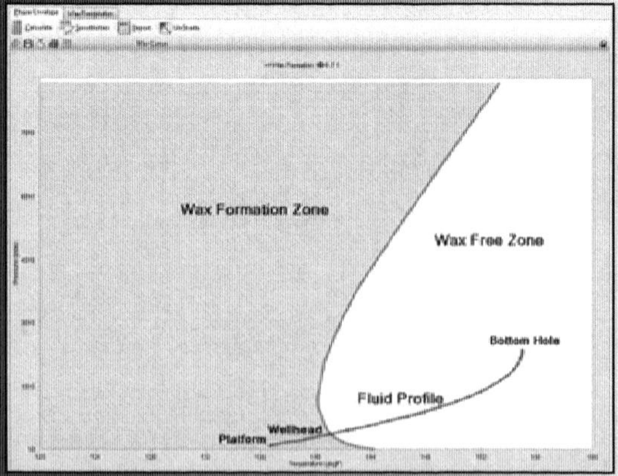

The plot on the left shows the predicted pressure – temperature profile of the fluids from bottom-hole to surface, showing how the wax-producing zone is entered toward the top of the wellbore.

Adding insulation to the design enables this profile to avoid the wax formation zone as shown in the plot on the right.

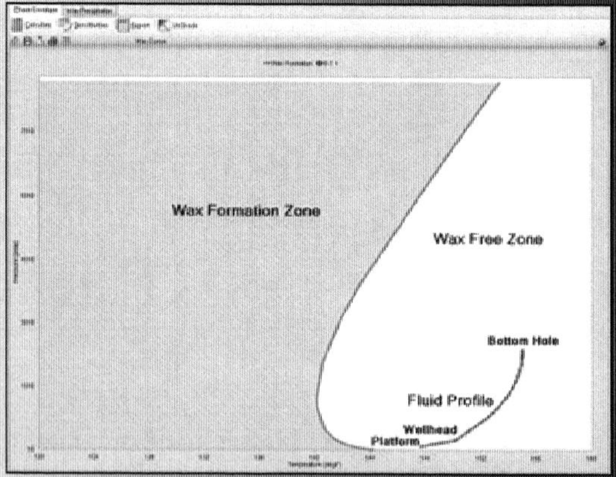

Interaction of Oil and Gas Shale with Drilling Muds and Fracturing Fluids

Samyukta Koteeswaran, Peter E. Clark

Department of Chemical Engineering, Oklahoma State University

Introduction

The boom in drilling and completions in shale formations has been partially responsible for the large increase in oil production in the United States [1]. To a large extent, development of shale resources has been facilitated by the use of high-rate fracturing treatments. These treatments typically use friction reducers to lower surface pressure requirements, and thus job costs. Probably the most common polymers used are polyacrylamides. While this family of polymers provides excellent friction reduction, they often interact with shales in a manner that can potentially cause formation damage problems. The polymer – shale interactions are well known and have been used for years to stabilize shales during drilling. Since shales typically have low permeability, it is difficult to ascertain the magnitude of formation damage by polyacrylamide based fracturing fluids.

The alteration of rock properties such as wettability and permeability is a common drilling problem [2]. Oil production will be restricted if the permeability is lowered. A change in wettability will have serious implications for both the core analysis and for formation damage, wettability also has a major impact in multi-phase flow in porous media [2]. The components that are blended with drilling mud are acceptable if they do not alter the rock properties. However, all the additives alter wettability and significantly reduce the permeability. Additives such as surfactants, clays, polymers and other chemicals could interact with the formation and could cause formation damage. Formation damage is caused by invasion of foreign fluids near the well bore [3, 4]. The study of rock mechanics and chemical interactions between shale and drilling mud is important before pumping drilling mud into the well as drilling fluids could considerably reduce the permeability [5].

Inhibited fluid systems that use polyacrylamides have proven effective for drilling through shale. Good shale inhibitors limit ion exchange into interlayer areas and prevent hydration of surfaces. The additives are adsorbed to the clay surface by hydrogen bonding to raise the zeta potential and thereby strengthen the clay gel structure [6, 7]. Shale stability can be achieved by combining potassium chloride with partially hydrolyzed polyacrylamides [8]. Potassium chloride is added to the mud system to reduce the swelling tendency of water sensitive shales.

Shales in each region behave differently when exposed to drilling fluid, polyacrylamides in particular. The interaction of the polyacrylamides with shales depends on number of factors such as swellability, permeability, surface charge, porosity and mineralogy of shales. The shale should be analyzed before injecting it with the drilling fluid. The mineral composition of the shales in this study is determined using SEM - EDS techniques. Surface characteristics and surface charge of the shale is analyzed using Scanning Electron Microscopy and Zeta Potential Analysis respectively. The interaction of polyacrylamides with the shales is quantified by measuring the rheology of shale slurries with polyacrylamides.

In this study the interactivity of shale with polyacrylamides is quantified and it is correlated to the surface properties of the shale.

Materials and Methods

Sample Preparation

Polymer

Anionic and Cationic Polyacrylamide (PAM) of average molecular weight of 10^7 g/gmol is used. Kemira supplied the polyacrylamide samples. The sample is measure by weight and added to DI water and left in shaker table at a speed of 200 RPM for 15 minutes. The sample is then left to hydrate at room temperature for 24 h.

Shale Sample

New Albany, Chattanooga, Gorgas and Woodford shale samples are cut using a coresaw from the core samples. The shale is ground using a belart mixer and sieved to obtain particles less than 75 micrometer. These samples are measured by weight and added to DI water and left in shaker table for 30 minutes at speed of 500 RPM. The sample is left to hydrate for 24 h at room temperature before adding it to the polymer sample.

Equipment

Rheometer

The Discovery HR-3 (DHR-3), advanced single head controlled stress rheometer is used for making rheological measurements of the sample. Rheometer is a highly sensitive device for probing molecular structure. Parallel plate geometry with gap from 0.5 mm to 1 mm is used for samples with viscosities >10 mPa.s. Double wall concentric cylinder is used for measuring the properties of low viscous samples. All the experiments were done at a temperature of 25 °C ± 0.03 °C. Three types of test were conducted on each sample. They are flow ramp test, frequency sweep, and amplitude sweep. Flow ramp tests are done for shear rates from 0.001 s^{-1} to 200 s^{-1}. The time is chosen depending on the viscosity of the samples. Frequency sweep test is done from 0.1 rad/s to 20 rad/s at 5% strain. Amplitude sweep is done from 0.1 – 100% strain at 10 rad/s. Pre shearing of all the samples for 30 s at 200 s^{-1} is done before each test. Amplitude sweep test is done to determine the strain % for the frequency sweep test.

Geometry

To make rheological measurements of the slurries vane geometry is used. The use of vane technique prevents wall slippage and also particle settling during the measurements.

Zeta Potential Analyzer

Zeta PALS model made by Brookhaven Instruments Corporation is used for measuring the net surface charge of the shale particles. The experiment was conducted and room temperature and repeated thrice to get concurrence.

Analysis

Model

Carreau Model was used to determine the zero shear rate viscosity for the entire polymer – shale and polymer samples. This five-parameter model has sufficient flexibility to fit a wide variety of experimental curves.

This viscosity model allows data to be fitted to the following model

$$\frac{\eta - \eta_\infty}{\eta_0 - \eta_\infty} = \frac{1}{(1 + (K_{1\gamma})^2)^{n^{1/2}}}$$

Where η_0 the Newtonian viscosity, η_∞ the infinite viscosity, K the consistency and n the power law index.

This model is used to fit data from a fluid that have a lower Newtonian and upper Newtonian region with a power law region in between. The Carreau model is a variant to Ellis model and an alternative to the Cross model. The Carreau model attempts to describe a wide range of fluids by the establishment of a curve-fit to piece together functions for both non-newtonian and shear thinning ($\eta < 1$) non-Newtonian laws.

Rheology Tests

Flow curve and Step Shear Rate

The polyacrylamide – shale sample is subject to continuous step shear rate from 0.001 s^{-1} to 200 s^{-1} . The flow ramp test is repeated thrice to get good concurrence. Flow ramp test is used to study the behavior of the material over a spectrum of shear rates. The zero shear viscosity of all the samples is determined by fitting the Carreau model using the inbuilt TRIOS software.

Scanning Electron Microscopy

SEM techniques were used to study the surface properties and morphology of the shale under study. Shale core were core sawed to get 1 – 2 mm shale particles, parallel to the bedding plane. The sample was placed on the stub and the shales being non conducting materials were sputter coated with conducting layers of carbon. The surface of the shale was examined using different magnifications. In order to determine the elemental composition of shales, the shale samples were coated with layers of carbon and Energy Dispersive Spectroscopy (EDS) analysis of the shale samples were done using Evex nanoanalysis software in SEM.

Results and Discussion

In this section we would be discussing the results of the rheology of shale samples with anionic and cationic polyacrylamide. The shale mineralogy, surface properties of shale and interaction of shale with the polyacrylamides is measured quantitatively. Table 1. Shows the mineralogy of shales under study.

Table 1: Mineralogy of shale samples determined using SEM – EDS

Elements	**New Albany (Wt. %)**	**Gorgas (Wt. %)**	**Chattanooga (Wt. %)**	**Woodford (Wt. %)**
Al	13.43	15.49	11.21	15.59
Si	27.39	58.98	76.71	67.38
Ti	0.74	1.4	1.13	-
S	19.77	-	0.36	-
K	9	12.76	7.95	10.54
Fe	29.68	11.36	2.63	4.3
Mg	-	-	-	2.2

Table 2. Shows the surface charge for each shale determined using the Zeta Potential analysis. It is clear that the net surface charge in all the shales under study is highly anionic in nature. The surface charge of the shale does not have a great influence on the interaction of polyacrylamide with shales but higher the net negative charge of shale lower is its interaction with anionic polyacrylamide.

Table 1: Zeta Potential of Shales

Shale	**Zeta Potential (mV)**	**pH**
New Albany	-28.45	8.12
Gorgas	-34.39	8.73
Chattanooga	-23.69	7.48
Woodford	-28.58	7.78

The interaction of polyacrylamide with shale is measured rheologically. In this method, the concentration of the shale is kept constant at 0.5 lb/bbl and the anionic polyacrylamide concentration is varied from 0.1 wt. % to 0.3 wt. %. This is the concentration range within which the polyacryalmides are used in drilling muds. Figure 1. Shows the zero shear rate viscosity change of anionic polyacrylamide – shale slurry for increasing concentration of anionic polyacrylamide. From the figure it could be seen that Chattanooga shale has a bigger increase in viscosity with increasing anionic polyacrylamide concentration and Gorgas shale has the least increase in viscosity. This clearly shows that each shale interacts differently with polyacrylamides.

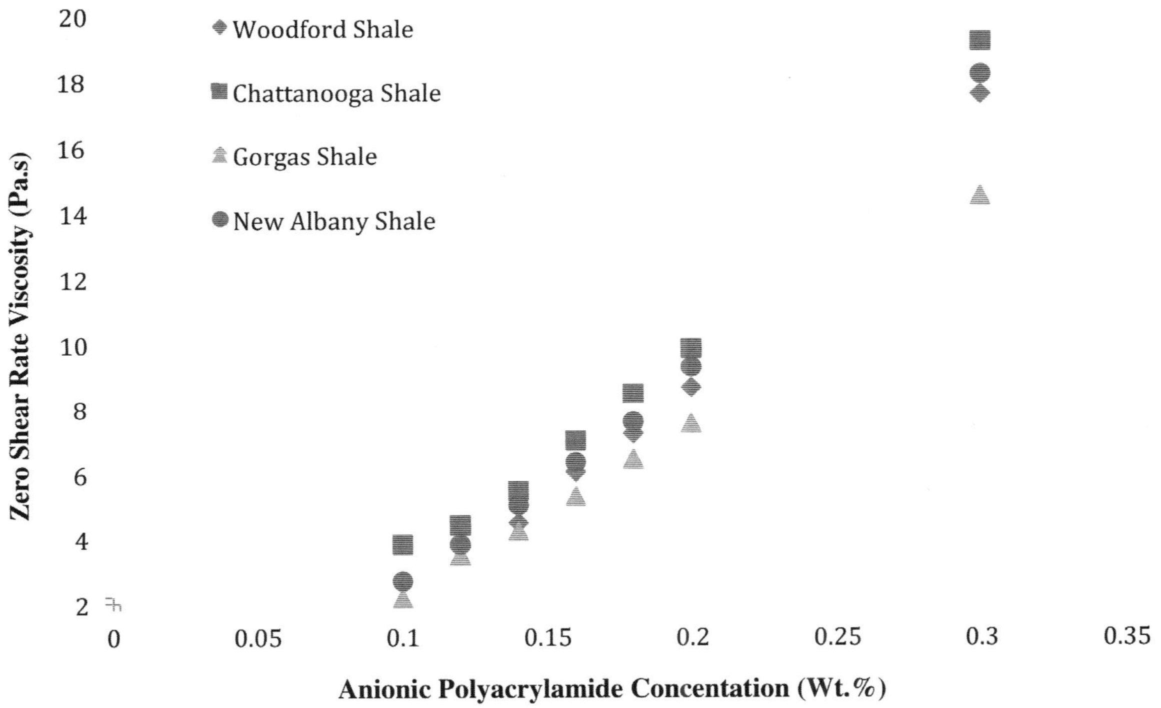

Figure 1: Zero Shear Rate Viscosity (Pa.s) vs Anionic Polyacrylamide Concentration (Wt. %) of the shale samples

Conclusion

Polyacrylamides are most commonly used friction reducers and the interaction of polyacrylamide shale interaction is not widely studied. In this paper, we have proposed a method to quantitatively measure the polyacrylamide – shale interaction. Each shale interacts differently with anionic polyacrylamide, with Chattanooga shale having highest interaction and Gorgas shale the lowest interaction with anionic polyacrylamide among the four shales under study. It is imperative that the shale – fluid interaction is studied for each shale formation before injecting the fluid.

Acknowledgements

We would like to thank the following organizations and people who helped us during the course of work

- Our funding source, Research Partnership to Secure Energy for America.
- Dr. Jack C. Pashin, School of Geology, Oklahoma State University.
- Dr. Jim Puckette, School of Geology, Oklahoma State University.
- Dr. Josh Ramsay, School of Chemical Engineering, Oklahoma State University.
- Department of Chemical Engineering, Oklahoma State University.

References

1. Shah, S.N., N. Shanker, and C. Ogugbue. *Future challenges of drilling fluids and their rheological measurements*. in *AADE fluids conference and exhibition, Houston, Texas*. 2010.
2. Sharma, M.M. and R.W. Wunderlich, *The Alteration of Rock Properties Due to Interactions With Drilling Fluid Components*. Society of Petroleum Engineers.
3. Mahto, V. and V.P. Sharma, *Rheological study of a water based oil well drilling fluid.* Journal of Petroleum Science and Engineering, 2004. **45**(1–2): p. 123-128.
4. Urbissinova, T., J.J. Trivedi, and E. Kuru, *Effect of Elasticity during Viscoelastic Polymer Flooding - A Possible Mechanism of Increasing the Sweep Efficiency*. Society of Petroleum Engineers.
5. Corr, C.C. a, and R.S.V. Nascimento, *Study of shale-fluid interactions using thermogravimetry.* Journal of Thermal Analysis and Calorimetry, 2005. **79**(2): p. 295-298.
6. Guo, W.-Y. and B. Peng, *Synthesis and characterization of starch-graft-polyacrylamide copolymers and their application as filtration control agents in drilling fluid.* Journal of Vinyl and Additive Technology, 2012. **18**(4): p. 261-266.
7. Han, M., et al., *Laboratory Investigation of the Injectivity of Sulfonated Polyacrylamide Solutions into Carbonate Reservoir Rocks*. Society of Petroleum Engineers.
8. Clark, R.K., et al., *Polyacrylamide/Potassium-Chloride Mud for Drilling Water-Sensitive Shales.*

400443 Evaluation of the Methods for Predicting Dynamic Viscosity of High-Temperatures and High-Pressures Deep Petroleum Formations

Abstract:

Ten methods of predicting dynamic viscosity for high-temperatures and high-pressures (HTHP) deep-water petroleum formations are evaluated for ease of implementation and applications for flow assurance modeling, pipeline flow simulation and reservoir simulation of carbon dioxide sequestration in the depleted oil-reservoirs. The evaluated techniques include various versions of the *dead-oil viscosity correlation*, the *Stiel-Thodos viscosity correlation*, the *Little-Kennedy correlation*, the *free-volume viscosity theory*, the *frictional theory for viscosity*, the *kinetic theory description for viscosity*, the *Van der Waals transport equations-of-state*, the *Ely-Hanley corresponding states viscosity methods*, the *hard-sphere theory of viscosity by Dymond-Assael*, the *Yarranton-Satyro viscosity correlations* and those versions of viscosity methods that are coupled with the modified Tait equation, VDW EoS, CPA EoS and SAFT+ cubic EoS for liquid density estimations at the HTHP conditions. The HTHP dynamic viscosity data used in the evaluation of the methods for predicting dynamic viscosities are reported in the literature by Enick Group at U of Pitt, Caudwell-Trusler-Wakeham Group at Imperial College London and Ducoulombier et al. in France. While all the evaluated methods gave various degrees of uncertainties at HTHP conditions (due to the judicious choice of the inaccurate liquid density correlations), the VDW viscosity equation of state (which does not require liquid densities or volumetric properties as input data) performs better and in agreement with the measured dynamic viscosities at the HTHP conditions of deep-water petroleum formations. It also provides a very suitable mean of predicting viscosity over the entire PVT states of HTHP densities, including dilute, dense-fluids, supercritical, near-critical and at critical viscosity. This article is useful for all type of analyses of flow assurance modeling.

Keywords: cubic equation, viscosity methods, dynamic viscosity, volumetric property, flow assurance

400463 The Van Der Waals Ten Commandments for Cubic Equations-of-State (Electronic Poster Presentation)

Abstract:

The reading of the Van der Waals (VDW) 1910 Nobel Prize Lecture[1,2] reveals the hindsight and foresights of the VDW theory of cubic equations of state for the individual pure substances and mixtures. A correct reading of the VDW Nobel Prize Lecture reveals some Do's and Don'ts and their consequences led the author to the following itemized Van der Waals Ten Commandments for Cubic Equations-of-State, which begins with Oh the Faithful:

- Be Loyal to the VDW Cubic Form of Attractive and Repulsive Expressions
- Be Loyal to the VDW Ultimate Objective: Construct A Substance-Based Cubic Equation
- Be Loyal to the VDW Asymptotic Critical Volume-Limit: Weak-Point of VDW Theory
- Be Loyal to the Four Properties of VDW Theory of Cubic Equations of State
- Be Loyal to the VDW Empirically-Based Molecular Parameters for Reformed Equations
- Be Loyal to the VDW Critical Point as Limit of Phenomenological Gas-Liquid Transition
- Be Loyal to the VDW Defined Gas-Constant for Unifying Caloric and Transport EOS

- Be Loyal to the VDW Continuity of Gas and Liquid States for Temperature Functions
- Be Loyal to the VDW Constraining Temperature Parameters to a(T) and b(T)
- Be Loyal to the VDW Corresponding States Principles for Generalized Property Charts

1. INTRODUCTION AND OBJECTIVE

We recently celebrated the hundred year anniversary[1,2] of the Nobel Prize Lecture delivered by J. D. Van der Waals on December 10, 1910 entitled "*The Equation of State for Gases and Liquids*."

This poster highlights the salient points, including the obvious and non-obvious facts about the author's reading of the 1910 Nobel Prize Lecture. In particular, areas of the Nobel Lecture where noticeable quotes have emerged over the years into textbooks, seminars, conferences and symposium are addressed from the viewpoint of the author's experience. The objective of the poster is to highlight fact that can be applied in the future development of the Van der Waals cubic equations of state.

2. DISCUSSION OF THE TEN COMMANDMENTS

It is revealed in this poster that the difficulties encountered with the modified VDW type of cubic equations can be traced to the violation of one or more of those VDW Commandments: as for instance the lack of *continuously* differentiable temperature functions a(T) and b(T) leads to difficulties in the supercritical regions or to the infinite isochoric heat capacity at the critical point or to the impossibility of simultaneously predicting accurate vapor pressure and virial coefficients with the same functional form of temperature-dependent parameters.

Applying the VDW gas constant, R^{VDW} provides a way to unite thermodynamic and transport equations of state and also it is a way for the direct use of the VDW cubic equations of state for predicting thermodynamic properties as opposed to the widely accepted route of the PVT derivatives: after all the PVT derivatives are based upon the temperature functions and ***not*** the VDW parameters a (T_c) and b (T_c).

Also, it is by using empirically-based physical parameters in the reformed VDW 1873 equation that would stop the proliferation of the multiplicity of 2-P and 3-P cubic equations of state. VDW theory is a modified form of the ideal gas-law with molecular-based empirical parameters for correcting the incomplete VDW theory.

Furthermore, those investigators that are passionately advocating the replacement of VDW repulsive term forget that the ***theory of liquids*** of the VDW theory is based solely on the VDW repulsion term as shown by the following limiting behavior:

$$\lim_{\substack{P \to \infty \\ \text{for all T}}} v \to b(T_c)$$

But, in accordance with the VDW theory, the Z_c of cubic equation ***solely*** depends on the ratio of b/v_c, as shown by

$$Z_c = \frac{1 - 2\left(\frac{b}{v_c}\right)}{2\left[1 - \left(\frac{b}{v_c}\right)\right]^2}$$

Perhaps knowing that fact, VDW stated in the 1910 Nobel Prize Lecture[2]: "the weak point of my theory is b," so whenever $b/v_c = 1/3$ (or $v_c = 3b$), $Z_c = 3/8$ of the VDW 1873 cubic equation of state.

By the way, before we agree with the notion that the VDW theory shows ***anomaly*** behavior at the fluid critical point, we need to examine the adherence to the following limiting critical behavior that should be imposed on all the VDW cubic equations of state:

$$\lim_{\substack{T\to T_c\\ P\to P_c}} v \to v_c \quad \text{and} \quad \lim_{\substack{T\to T_c\\ P\to P_c}} Z \to Z_c$$

That stipulated physical boundary condition states that the VDW cubic equations of state should predict at fluid critical point: accurate ***critical volume*** and ***critical compressibility factor*** of the individual pure substances or mixture of fixed overall composition. In that case, the parameters of cubic equations of state should be constrained by ***four*** critical constraint criteria because there are *four properties* of the VDW 1873 equation: Z_c =3/8; Ω_a = 27/64; Ω_b = 1/8; Ω_w (or b/v_c) = 1/3 and that is in accord with the theory of cubic polynomial equations. Thus, *four unrelated* parameters are required in the VDW 1873 cubic equation; simply, parameters a and b are necessary but ***insufficient*** to resolve the VDW 1873 cubic equation at fluid critical point); the required *four* critical constraint criteria expressions are stated as:

$$\left(\frac{\partial P}{\partial v}\right)_T = 0; \quad \left(\frac{\partial^2 P}{\partial v^2}\right)_T = 0$$

$$f(P_c, v_c, T_c) = 0; \quad (v - v_c)^3 \equiv (Z - Z_c)^3 = 0$$

Besides the combining rules for parameters a_m and b_m that VDW introduced in 1888, the justification for more composition-dependent parameters in the VDW 1873 cubic equation is based on the *insufficiency* of the expressions for the critical properties:

$$P_c = \frac{1}{27}\frac{a_m}{b_m^2}; \quad T_c = \frac{8}{27}\frac{a_m}{Rb_m}; \quad v_c = 3b_m$$

While P_c, v_c and T_c vary with composition of mixture in the expression for the critical properties, the Z_c does not vary with composition, being identically the same as for the pure substance; that is the major source of errors in predicting critical properties from the VDW 1873 equation. Consequently, additional parameters are required to be embedded in the reformed VDW 1873 equation to express $Z_c = f$ (mixture composition). That is another justification for the design of the four-parameter Lawal-Lake-Silberberg equation of state.

The statistical mechanical derivation of the virial coefficients from the VDW 1873 equation is reported by Hill (1947, 1948)[3] as inaccurate after the second virial coefficient; as seen from the virial expansion of the VDW 1873 cubic equation,

$$Z = 1 + \left(b(T_c) - \frac{a(T_c)}{RT}\right)\frac{1}{v} + \left(\frac{b(T_c)}{v}\right)^2 + \left(\frac{b(T_c)}{v}\right)^3 + \left(\frac{b(T_c)}{v}\right)^4 + \cdots$$

The virial expression shows that the attractive parameter a (T_c) ***disappears*** from the *third and higher* virial coefficients: that is ***not*** VDW theory because the VDW concept is $Z_{rep} + Z_{att}$. Therefore, the remedy is for more parameters in denominator of a/v^2 term (as the theory of cubic polynomial equations stipulates *four unrelated* parameters for cubic equations), which was understood by Clausius (1880) and Berthelot (1900) in their reformed VDW cubic equations of state. However, it is not always obvious from the coefficients of the virial expansion of the VDW 1873 equation that the second virial is the *building blocks* for higher virial coefficients; the fact is shown by the following limiting values of PVT data for the second (B), third (C) and fourth (D) virial coefficients:

$$B = \lim_{1/v\to 0} (Z - 1)v$$

$$C = \lim_{1/v\to 0} [(Z - 1)v - B]v$$

$$D = \lim_{1/v\to 0} [(Z - 1)v^2 - Bv - C]v$$

Those limiting values of PVT data uniquely justified more parameters in the denominator of the a/v^2 term of the VDW 1873 equation of state because the expression of the second virial coefficient (B) requires parameters $a(T_c)$ and $b(T_c)$ as the building blocks for the third and higher virial coefficients: that is another justification for the construction of the four-parameter Lawal-Lake-Silberberg equation of state.

By predicting physical properties using the VDW theory of cubic equations of state, the author agrees with the statement by Guggenheim (**1945**)[4] that the corresponding states principle "may safely be regarded as the most useful by-product of the Van der Waals 1873 equation of state." Therefore, besides the temperature-dependent parameters a (T) and b (T), no other parameters embedded in the reformed VDW cubic equations of state should depend on temperature, otherwise the *meaning* of the Law of Corresponding States (LCS) is ambiguous. Hence, temperature-dependent binary interaction parameter is meaningless in the context of the LCS. The design of cubic equations of state reported by Himpan (1951) and Heyen (1980, 1981), having more temperature-dependent parameters than stipulated by the LCS, violates the corresponding states principle.

3. CONCLUSION

By adherence to the unspoken words of Van der Waals, which are casted as Ten Commandments in this poster, we can resolve the major disagreement between the Van der Waals theory and fluid properties, including the fluid critical point. Evidently, there should be a particular reason to construct another VDW cubic equation of state, otherwise we would reached the point of diminishing returns and joint others in the high degree of trivialization of the VDW theory of cubic equations of state.

REFERENCES:

1. **Journal of Supercritical Fluids,** Volume 55, Issue 2, Pages 401-860 (December 2010)

100th year Anniversary of van der Waals' Nobel Lecture (Edited by Sona Raeissi, Maaike C. Kroon and Cor J. Peters)

2. Van der Waals, J. D., **1910** Nobel Prize Lecture available from "J. D. van der Waals - Nobel Lecture: The Equation of State for Gases and Liquids". http://www.nobelprize.org/nobel_prizes/physics/laureates/1910/waals-lecture.html

3. Hill, T. L., "Free-Volume Models for Liquids," J. *Phys. and Colloid Chem.*, *51*, 1219 **1947**; "Derivation of the Complete Van der Waals' Equation from Statistical Mechanics," *J. Chem. Educ.*, *25* (6), 347, **1948**

4. Guggenheim, E. A., The Principle of Corresponding States. *J. Chem. Phys*. 13, 253, **1945**

400466 Characterization of Critical Properties and Acentric Factors for High-Molecular-Weight-Fluids, Petroleum Fractions, Coal and Natural Products

Abstract:

Characterizing methods reported by Walas (1985), Poling et al. (2001) and Ihmels-Ambrose (2010) for pure substances are based on structural formula and therefore not suitable for pseudo-component properties (or poorly characterized mixtures encountered with petroleum, coal and natural products, which are usually based on measured properties such as average molecular weight, boiling point and specific gravity (at 20^oC)[1] rather than molecular structure). The Paraffinic-Naphthenic-Aromatic (PNA) reported by Peng- Robinson (1978) or the group contribution methods by Lydersen et al. (1955) (or the methods based on carbon number reported by Whitson (1983) and Panuganti-Chapman (2013) are not considered in this report because the intention here is not to gain molecular insights into crude-oil properties but rather to treat pseudo-component properties as *single-lumped component* for the sole purpose of modeling by using cubic equations of state. Also, the trusted Database as tabulated by the NIST Groups for the critical properties and acentric factors do not include n-tetracontane ($nC_{40}H_{82}$), even though specific volumes at several temperatures and pressures for n-tetracontane have been reported by Arthur Doolittle since 1964 (*J. Chem. Eng. Data*, *9* (2), 275, 1964).

Consequently, thirty-six methods of characterizing critical properties and acentric factors for the parameters of cubic equations of state used in the process and reservoir simulation are evaluated for ease of implementation and use for high-molecular-weight-fluids, petroleum fractions, coal tar liquids and natural products. Analyses of results shows the methods are not adequate for poorly characterized pseudo-components of high aromatic, naphthenic and cycloalkanes;

and beyond n-Octadecane hydrocarbons, some critical property parameters (critical pressure and volume, critical-Z factor and acentric factor: which is greater than 1 and therefore, is physically meaningless) are usually based or trial-and trial error techniques: thus, lacking consistency with lighter hydrocarbons in the petroleum fractions assay. While most of the estimation methods provide satisfactory predictions of the critical temperatures, they all underestimated critical pressures of the pseudo-components.

By using the API Database as reported by the University of Pennsylvania Group and gathering data from *Hydrocarbon Processing* and other relevant sources (theses and dissertations reports), we developed a Paraffinic-Naphthenic-Aromatic Characterization Factor into the correlation for critical temperature in terms of average molecular weight (Mw), boiling point (T_b) and specific gravity (S_g) and we use several cross-plotting of the ratio of boiling point to critical temperature (T_b/T_c) versus molecular weight, critical properties (P_c, V_c) and acentric factor (ω) to generate high-correlation coefficients; correlating groups, such as (T_b/T_c), (T_c/P_c), (V_c/Z_c), (M_w/P_c), (M_w/T_c) and Z_c versus molecular weight or acentric factor show a high-correlation coefficients. While we claim no theoretical basis (apart from the high-correlation coefficients and the statistically derived predictive groups), the principles of dimensional similitude provide the convenient basis in the selected groupings. The grouping parameters reduce the number of tailored-fitting parameters and thus preclude the need for extensive experimentally measured data for the unstable petroleum fractions and pseudo-components.

Hundred petroleum fractions reported by Roess (1936), Lenoir-Hipkin enthalpy (1973) and petroleum fraction data are used to validate the pseudo-component critical properties developed in this reported. In comparison with available correlative methods, the current approach is better in accuracy and ease of implementation in cubic equations of state modeling. The current approach is also validated for predicting viscosity of coal liquids (Baltatu, 1985; 1995) as well as predicting densities of natural products. While the current approach may not be convenient for the type of work in the petroleum refining operations, the methodology is superior for empirical characterization of pseudo-component critical properties and acentric factors as lumped-single component and it thus offers another technique to the traditional *pseudo-component*, or *hypothetical-component* modeling approach for process and reservoir simulation models. It is therefore recommended for high-molecular-weight components for which experimental critical properties are typically unavailable.

References:

1. Riazi, R. M. *Characterization and Properties of Petroleum Fractions*; ASTM: Philadelphia, **2005**

2. API, *Properties of Hydrocarbons of High Molecular Weight*, API Research Project 42, American Petroleum Institute, Washington, DC, **1966**.

3. Riazi, M. R., Daubert, T. E., "Simplify Property Prediction," *Hydrocarbon Process, Petrol. Refiner*, *59*(3), 115-116, **1980**

4. Lin, H. M., Chao, K. C., "Correlation of Critical Properties and Acentric Factor of Hydrocarbon Derivatives" *AIChE J.*, *30*(6), 153-158, **1984**

5. Watson, K. M., Nelson, E. F., Murphy, G. B., "Characterization of Petroleum Fractions," *Ind. Eng. Chem. 27*, 1460-4, **1935**.

6. Winn, F. W., "Simplified Monograph Presentation, Characterization of Petroleum Fraction," *Pet. Refin.*, *36*(2), 157, **1957**

7. Smith, R. L., Watson, K. M., "Boiling Points and Critical Properties of Hydrocarbon Mixtures," *Ind. Eng. Chem.*, 29, 1408, **1937**

8. Watansiri, S., Owens, V. H., Starling, K. E., "Correlation for Estimating Critical Constants, Acentric Factor, and Dipole Moment for Undefined Coal-Fluid Fractions," *Ind. Eng. Chem. Process Des. Dev.*, *24*, 294-96, **1985**

9. Riazi, M. R., Daubert, T. E., "Characterization Parameters for Petroleum Fractions," *Ind. Eng. Chem. Res.*, *26*(24), 755-759, **1987**

10. Abbott, M. M., Kaufmann, T. G., Domash, L., "A Correlation for Petroleum Liquid Viscosities of Petroleum Fractions," *Can. J. Chem. Eng., 49,* 379, June, **1971**

11. Riazi, M. R., Al-Otaibi, G. N., "Estimation of Viscosity of Liquid Hydrocarbon Systems," *Fuel 80*, 27-32, **2001**

12. Cullick, A. S., Pebdani, F. N., Griewank, A. K., "Modified corresponding states method for predicting densities of petroleum reservoir fluids," *Ind. Eng. Chem. Res.*, *28* (3), 340-47, **1989**

13. Moharam, H. M., Fahim, M. A., "Prediction of Viscosity of Heavy Petroleum Fractions and Crude Oils Using a Corresponding States Method," Ind. *Eng. Chem. Res.*, *34* (11), 4140-44, **1995**

14. Sharma, R., Sunil Goel, S., "Correlation of Viscosity of Coal Liquids," *Ind. Eng. Chem. Res.*, 36 (9), 3999-4007, **1997**

15. Baltatu, M. E., "Prediction of the liquid viscosity for petroleum fractions," *Ind. Eng. Chem. Proc. Des. Dev.*, *21* (1), 192-5, **1982**

16. Baltatu, M. E., Ely, J. F., Hanley, H. J. M., "Thermal Conductivity of Coal-Derived Liquids and Petroleum Fractions," *Ind. Eng. Chem. Process Des. Dev.*, *24* (2), 325, **1985**

17. Baltatu, M. E., Chong, R. A., Huber, M. L., Laesecke, A., "Transport Properties of Petroleum Fractions," *Int. J. Thermophysics*, *20* (1), 85-95, **1999**

18. Jacoby, R. H., Yarborough, L., "PVT Measurements on Petroleum Reservoir Fluids and Their Uses," *Ind. Eng. Chem.*, *59*(10), 49, **1967**

19. Twu, C. H., "An Internally Consistent Correlations for Predicting the Critical Properties and Molecular Weights of Petroleum and Coal-Tar Liquids," *Fluid Phase Equil.*, *16*, 137, **1984**

20. Allen, J. C., "A Review of the Properties of Some Oil and Gas Reservoirs," *Oil and Gas J.*, *51*(12), 355, **1952**

21. Cragoe, C. S., "Thermodynamics of Petroleum Products," *Bureau of Standard, U. S. Department of Commerce, Miscellaneous Publication*, No. 97, 22, **1929**

22. Roess, L. C., "Determination of Critical Temperature and Critical Pressure of Petroleum Fractions by a Flow Method," *J. Soc. Pet. Tech.* (London), *22*, 665, **1937**

23. Lenoir, J. M., Hipkin, H. G., "Measured enthalpies of eight hydrocarbon fractions," *J. Chem. Eng. Data*, *18* (2), 195-202, **1973**

24. Moharam, H. M., Braek, A. M., Fahim, M. A., "A Simple Method of Estimating the Enthalpy of Petroleum Fractions," *Ind. Eng. Chem. Res.*, *37* (12), 4898-4902, **1998**

25. Filipe A. Z., Da Silva Lopes, V., Kontogeorgis, G. M., "Approach Suitable for Screening Estimation Methods for Critical Properties of Heavy Compounds," *Ind. Eng. Chem. Res.*, *45* (1), 476-480, **2006**

26. Jianzhong, Z., Biao, Z., Suoqi, Z., Renan, W., Guanghua, Y., "Simplified Prediction of Critical Properties of Nonpolar Compounds, Petroleum, and Coal Liquid Fractions," *Ind. Eng. Chem. Res.*, *37* (5), 2059-60, **1998**

27. Brule, M. R., Lin, C. T., Lee, L. L., Starling, K. E., "Multiparameter corresponding-states correlation of coal-fluid thermodynamic properties," *AIChE J.*, *28* (4), 616-25, **1982**

28. Wilson, G. M., Johnston, R. H., Hwang, S. C., Tsonopoulos, C., "Volatility of coal liquids at high temperatures

and pressures," *Ind. Eng. Chem. Proc. Des. Dev.*, *20* (1), 94-104, **1981**

29. Tsonopoulos, C., Heidman, J. L., Hwang, S. C., *Thermodynamic and Transport Properties of Coal Liquids*, New York, **1986**

30. Whitson, C. H., "Characterizing hydrocarbon plus fractions," *Soc. Pet. Eng. J. 23*, 4, 683-694, **1983**

31. Panuganti, S. R., Vargas, F. M., Chapman, W. G., Property Scaling Relations for Nonpolar Hydrocarbons, *Ind. Eng. Chem. Res.*, **2013**, *52*, 8009-802

32. Lydersen, A. L., Greenkorn, R. A., Hougen, O. A., *Generalized Thermodynamic Properties of Pure Fluids*, University of Wisconsin, Engineering Experimental Station, Report 4, Madison, Wisconsin, October, **1955**

33. Lydersen, A. L., "Estimation of Critical Properties of Organic Compounds by the Method of Group Contributions," University of Wisconsin, Engineering Experimental Station, Report 3, Madison, Wisconsin, April, **1955**

34. Christian Ihmels, E., "The Critical Surface," *J. Chem. Eng. Data*, *55* (9), 3474-3480, **2010**

35. Sovova, H., Galushko, A. A., Stateva, R. P., Rochova, K., Sajfrtova, M., Bartlova, M., "Supercritical fluid extraction of minor components of vegetable oils: β-Sitosterol," *Journal of Food Engineering*, *101*(2), 201-209, **2010**

36. Freitas, S.V.D., E Silva, F.A., Pastoriza-Gallego, M. J., Piñeiro, M. M., Lima, A. S., Coutinho J. A. P., "Measurement and prediction of densities of vegetable oils at pressures up to 45 MPa," J. Chem. Eng. Data, 58 (11), 3046-53, **2013**

37. Poling, B. E., Prausnitz, J. M., O'Connell, J. P., *The Properties of Gases and Liquids*, (5th ed.) McGraw-Hill, New York, **2001**

38. Walas, S. M., *Phase Equilibria in Chemical Engineering*, Butterworth Publishers, Boston, **1985**

400526 Molecular Interpretation of Parameters in the Van der Waals Theory of Cubic Equations-of-State

Abstract:

Even though it is universally known that Van der Waals 1873 theory of cubic equations of state predicts inaccurate liquid-phase properties, unreliable for highly polar and hydrogen-bonded fluids, inappropriate for fluid critical point, unsuitable for large components and incapable of accurate virial coefficients, but as yet it is not widely stated that those fluid properties depend on the molecular structure of pure substances which is *evidently* not reflected by the parameters in the VDW theory. This poster suggested the possibility of including two inspired parameters into the Van der Waals 1873 equation of state. The physically-based empirical parameters that replicate molecular shapes and molecular structure of pure substances are introduced into the reformed Van der Waals (VDW) 1873 equation to reconcile the VDW 1873 theory with the stipulated shapes of the potential energy and length scale of the statistical mechanical potential functions.

By relating two of the empirical parameters (α, β) in the reformed VDW 1873 equation (which has been affectionately named the Lawal-Lake-Silberberg (LLS) cubic equation) to the microscopic force-parameters derived from the statistical mechanics energy potential functions and performing perturbation and other sensitivity analysis; the *sensitivity* of the derivatives of the structural parameters (P_c, Z_c, T_c) and the *sensitivity* of the derivatives of second, third and fourth virial coefficients with respect to α show non-linear trend with α-parameter and thus, α replicates *structure* of pure substances in the configuration of the LLS equation. Similar analysis of the derivatives of the critical volume (V_c) and Boyle temperature (T_B) with respect to β show non-linear trend with β-parameter and thus β reflects *shapes* of pure substances in the configuration of the LLS equation. Nonetheless, re-constructing the VDW 1873 cubic equation from the microscopic to the macroscopic level has been very challenging because the attraction term of the 1873 theory comes from the VDW intuition, as seen in the Schematic Figure. However, those empirical parameters reflecting structure and shape of pure substances in the LLS equation resolve the major disagreement between the VDW theory and fluid properties, including fluid critical point.

Since the shape-structure parameters (α, β) occur in the intuitive-part of the VDW theory, we can speculate that instinctive reasoning of Van der Waals is *correct* but rather Van der Waals *forgot* to carry his theory of cubic equations of state to logical conclusion before passing away in 1923. We can apply that inference to attest to the widely known but not often said that "*some pertinent variables must have been overlooked in the development of cubic equations of state*," which was echoed by Sydney Young (1890) in regard to the analysis of the coexistence gas-liquid volumes and later repeated by Stanley Walas (1985) in regard to the State of The Arts Review of cubic equations of state in his book: *Phase Equilibria in Chemical Engineering*. Stanley Walas' conclusion is "it does appear, however, that experimental data often are more nearly in accord with the *principle of corresponding states* than with specific EOS," which reinforced the statement by Guggenheim (1945) that the corresponding states principle "may safely be regarded as the most useful by-product of the Van der Waals 1873 equation of state."

The Schematic Figure is, however, contradictory to the widely accepted viewpoint of molecular theory in which repulsive forces are responsible for structure of liquids while the attraction forces play no role whatsoever in the theory of liquids. In contrast, the Schematic Figure illustrates that the attractive pressure term of the VDW equation is responsible for structure of liquids. Also, the Schematic Figure only assists in the understanding of the thermodynamic properties of dense fluids from macroscopic viewpoint *not* the interpretation of those properties in terms of the forces between the molecules. The conjecture of the Schematic Figure makes no claim whatsoever to quantify geometrical shape and structure of molecules (i.e. bonding distance and angles) which are in the realm of the applications of quantum chemistry as exemplified by the works of 1998 Nobel Laureates John Pople and Walter Kohn. Furthermore, the Schematic Figure cannot apply at molecular level to compute shapes of molecules because the shapes of Potential Energy Surface resulted from interplay of *four* basic interaction energies (electrostatic, exchange, induction, and dispersion) as discussed in *Chemical Reviews*.

While the shape-structure parameters are weighted by molecular size (or molecular volume) in the Schematic Figure, it is the *discriminant* of the attractive pressure term in the VDW equation that is responsible for accurate saturated liquid densities because $Z_c = f(\alpha, \beta)$. As typified by the coefficient of the logarithm term and the expression in the ln{} of the fugacity coefficient being depended on the discriminant, as $Z_c = f(\alpha, \beta)$. Since the Z_c factor solely depends on the limiting critical volume parameter Ω_w (or b/v_c) and they both can be expressed as $f(\alpha, \beta)$, they should be specified individually in cubic equations of state. In that regard, however, we may have unintentionally addressed the lament of Van der Waals which he stated in the 1910 Nobel Prize Lecture[1-2] as "*And here I have come to the weak point in the study of the equation of state. I still wonder whether there is a better way. In fact this question continually obsesses me, I can never free myself from it, and it is with me even in my dreams*." A better way is to use Z_c and Ω_w to specify all the parameters of cubic equations of state because Z_c solely depends on Ω_w and that would fulfil the VDW stated objective[2]: "*to determine the relation between p, v and T for a substance.*"

400545 Rationale for the Parameterization of Van Der Waals Equations-of-State: The Lawal-Lake-Silberberg Equation As Classic Prototype

Abstract:

As of yet and by judging from the State-of-The-Arts Reviews by Leland (1968), Tsonopoulos-Prausnitz (1969), Vera-Prausnitz (1972), Abbott (1973), Storvick-Sandler (1977), Horvath (1974), Prausnitz (1977), Firoozabadi (1978), Martins (1979), Wong-Prausnitz (1985), Walas (1985), Anderko (1990), Wu-Prausnitz (1998), Poling et al. (2001), Wang-Pope (2002), Valderrama (2003) and Prausnitz -Tavares (2004), the issue of which Van der Waals (VDW) model with the same number of parameters is best does not have a universal and widely acceptable answer and there is still no obvious choice among the three- and four-parameter cubic equation of state models.

This contribution asserts the reformed four-parameter VDW 1873 equation of state, which has been affectionately named the Lawal-Lake-Silberberg (LLS) cubic equation, as the preferred choice among all categories of the VDW type of cubic equations of state by honoring the *four* properties of the VDW theory, the *four* critical constraint criteria and by being in accord with the theory of cubic polynomial equations of state which stipulates four-parameter for all cubic polynomial equations. A theoretical justification within the framework of the VDW empiricisms is provided for the LLS equation as the Substance-Based Cubic Equation of State for the individual pure substances, in agreement with the VDW objective which he stated toward the end of the 1910 Nobel Prize Lecture*** as "*to determine the relation between p, v and T for a substance*." However, we are still lacking the microscopic-level description for the two empirical parameters introduced into the reformed VDW 1873 equation, the LLS equation.

The LLS equation is applicable to all categories of pure substances, regardless of the chemical structure, polarity or molecular size and prediction of accurate volumetric and thermodynamic properties over the entire PVT states are guaranteed. But, the extension of the combining mixture rules to ternary and multicomponent systems is still being explored.

*** Van der Waals, J. D., **1910** Nobel Prize Lecture available from "J. D. van der Waals - Nobel Lecture: The Equation of State for Gases and Liquids". http://www.nobelprize.org/nobel_prizes/physics/laureates/1910/waals-lecture.html

400871 Application of Artificial Neural Networks to Modelling Vapour–Liquid Equilibrium in the Fischer–Tropsch Reaction

Abstract:

Modelling vapour–liquid equilibrium (VLE) requires a choice of the equation of state (EOS), the choice of pure component parameters, and the choice of mixing rules. The more constants the EOS has, the more mixing rules are required, and the more data is required for evaluating pure-component parameters. The Fischer–Tropsch (FT) reaction mechanism has been studied for over 90 years yet an adequate description of its kinetics still seems to escape us. Including VLE in modelling an FT reactor only compounds the intricacy.

The field of computational intelligence is revolutionizing many areas of science and engineering. The ability to extract useful insights from vast amounts of weak and incomplete information is not only fuelling the current interest in "big data", but has also enabled rapid progress in more traditional disciplines such as computer vision, estimation, and robotics, where data are available but difficult to interpret.

The FT synthesis face similar challenges, as many molecular and reaction pathways are still poorly characterized and available data are incoherent. Therefore, machine learning techniques such as artificial neural networks may be instrumental in overcoming these challenges by modelling the dependencies between random variables and using them to extract and accumulate the small amounts of information each random event provides. Hence, this study applies artificial neural networks to the complex problem of modelling VLE in the FT reactor. The data used to train, validate, and test the model is obtained from literature.

Role of Surface Chemistry in Predicting Viscosity of Concentrated Carbonaceous Solid-Water Slurries

Amrita Mukherjee, Sarma V. Pisupati#
John and Willie Leone Department of Energy and Mineral Engineering and the EMS Energy Institute,
110 Hosler Building,
The Pennsylvania State University, University Park PA 16802
#corresponding author: sxp17@psu.edu

Prepared for Presentation at
American Institute of Chemical Engineers
2015 Spring Meeting
11th Global Congress on Process Safety
Austin, Texas
April 27-30, 2015

Role of Surface Chemistry in Predicting Viscosity of Concentrated Carbonaceous Solid-Water Slurries

Amrita Mukherjee, Sarma V. Pisupati#
John and Willie Leone Department of Energy and Mineral Engineering and the EMS Energy Institute,
110 Hosler Building,
The Pennsylvania State University, University Park PA 16802
#corresponding author: sxp17@psu.edu

Keywords: viscosity, coal, slurry, surface chemistry, inter-particle interaction

Abstract

Viscosity prediction of carbonaceous solid-water slurries is essential to process design and control. Existing viscosity models take into account particle-size distributions but fail to account for the surface chemistry of the solids. Surface chemistry affects inter-particle and particle-water interactions and therefore influences slurry viscosity of concentrated suspensions. On the basis of surface chemistries characterized by contact angle and zeta potential measurements, this study determines the inter-particle interaction energies. Polar interaction energy, observed to be 2-3 orders of magnitude greater than the electrostatic interaction energy and the van der Waal interaction energy, is clearly the dominant interaction energy for such a system. Both hydrophobic and hydrophilic interactions of these particles in water result in microstructures which trap water either in the form of coalescing droplets in aggregation networks or in the form of hydration layers around carbonaceous solids, leading to an increase in the effective solid volume fraction and thus increase in slurry viscosity. The increase in effective solid volume fraction is dependent on the surface chemistry of the solid and thus is specific to a carbonaceous solid. The factor by which solid volume fraction increases for a particular carbonaceous solid was determined using viscosity measurements and Krieger-Dougherty equation and was found to correlate very well with the oxygen to carbon ratio of the solid. Incorporating this factor in the Krieger-Dougherty equation resulted in accurate prediction of slurry viscosity. This modified model was successfully validated using three other concentrated carbonaceous solid-water slurries.

1. Introduction

Carbonaceous solid-water slurries are concentrated suspensions of coal or petcoke particles in water and are used as gasifier feedstock [1][2]. A very high solids loading, in the range of 60-75% (by weight), results in an increased slurry viscosity [3]. The slurry viscosity determines the fluid flow behavior and the pumping energy requirement. Accurately determining slurry viscosity is a challenging task; but it is also vitally important. Solids loading in the slurries should be high for gasification efficiency but at the same time viscosity of the slurries should not be too high. Higher viscosity leads to difficulties in handling and higher pumping energy consumption. The challenge in determining slurry viscosity is complicated primarily due to variables such as different particle-size distributions and widely varying surface chemistry of the carbonaceous solids along with the additives used and pH of the slurry.

Existing models for predicting viscosities of suspensions are mainly semi-empirical equations predicting viscosity on the basis of solid volume fraction and particle-size distribution [5][6]. Since the models were originally developed for non-interacting hard spheres, these do not take into account surface chemistry of the solids [7]. Among these semi-empirical equations, Krieger-Dougherty equation was proved to be the most effective semi-empirical model and therefore has been used in this work [8][9][10]. Other models such as the phenomenological model developed by Dooher for viscosity prediction accounted for surface chemistry through statistical correlations [11]. Moreover their work only involved bituminous and sub-bituminous coal and did not include the entire range of carbonaceous solids. Usui proposed a thixotropic viscosity model which can take into account the agglomerative nature of the hydrophobic carbonaceous solids but is not applicable for those carbonaceous solids like low rank coals where slurry viscosity is affected due to adsorption or absorption of free water and not aggregation of solids [12][13]. Therefore, clearly there is need for a viscosity model which can take into account surface chemistries of the entire range of carbonaceous solids, along with particle-size distribution, to predict viscosity accurately.

Surface chemistry of coals depend on the rank of the coal. Low rank coals have higher oxygen, moisture and mineral matter content as compared to high rank coals, petcoke and bitumen. For the same particle-size distribution and same solids loading, these carbonaceous solid-water slurries have widely differing viscosities owing to the differences in their surface chemistries[14].

This study follows a fundamental approach to analyze the effect of surface chemistry on slurry viscosity. Different surface chemistries lead to different particle-particle and particle-medium interactions which in turn result in different suspension microstructures. These suspension microstructures cannot be studied using the traditional, highly sophisticated light based techniques because of the micron or even millimeter sized particles of solids in the slurries and the high concentration of the slurries. Therefore in order to determine the suspension microstructure, this study calculates the inter-particle interaction energies on the basis of contact angle and zeta potential measurements of the carbonaceous solids. On the basis of the understanding of the inter-particle interaction mechanism and thus the suspension microstructure, the semi-empirical Krieger-Dougherty equation is modified to account for the surface chemistry of the carbonaceous solid. The modified equation is validated using three other carbonaceous solid-water slurry samples and provides an accurate prediction of viscosity.

2. Experimental

2.1. Materials. Pust coal (lignite), Dietz coal (subbituminous), Illinois #6 coal (bituminous), Orchard coal (anthracite) and petcoke were ground in a laboratory-sized ball mill and were then screened and separated into the following size ranges: -250+212 μm, -212+180 μm, -180+150 μm, -150+125 μm, -125+106 μm, -106+90 μm, -90+53 μm, -53+44 μm, -44+20 μm, -20 μm. Experiments were carried out with slurries of these carbonaceous solids in water. The same particle-size distribution, as shown in figure 1(a) was used for all the experiments excepting the experiments concerning model validation. Particle-size distribution was maintained constant to highlight the effect of surface chemistry of the carbonaceous solids on slurry viscosity.

Model validation was performed using petcoke, Pittsburgh no. 8 (bituminous coal) and Beulah (lignite coal) coal-water slurries. Different particle-size distributions and two different carbonaceous solid samples were used to observe the sensitivity of the model to particle-size distribution and surface chemistry. Particle-size distributions for the three samples used for model validation are shown in figure 1(b). Proximate and ultimate analyses of all the samples are presented in table 1:

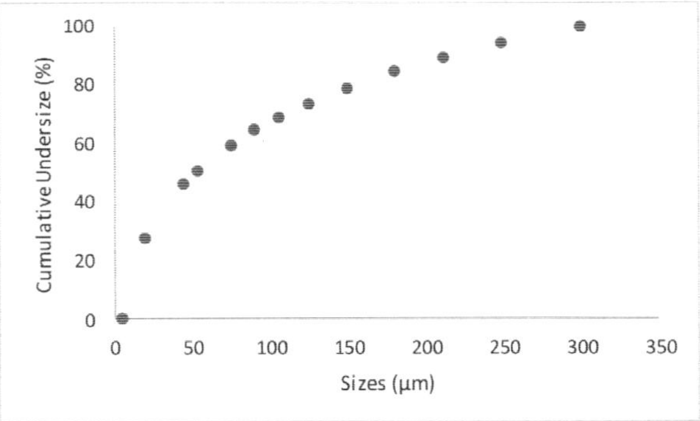

Fig 1 (a): Particle-size distribution of the carbonaceous solid particles in their respective slurries

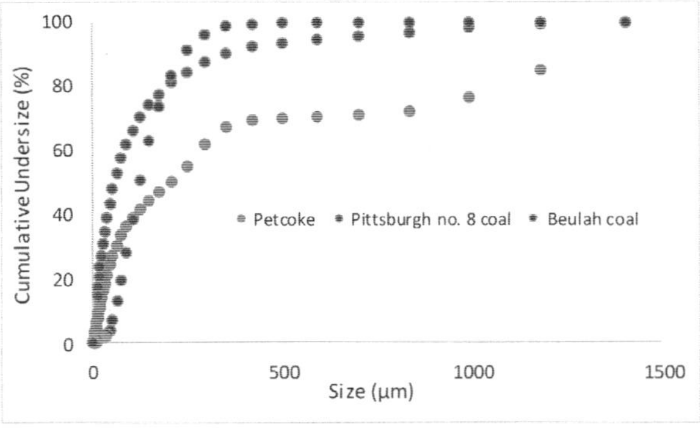

Fig 1 (b): Particle-size distribution of the carbonaceous solid particles in their respective slurries in the model validation experiments

Table 1: Proximate and ultimate analyses results:

	Proximate Analysis (weight %, as received basis)				Ultimate Analysis (weight %, as received basis)				
	M	VM	A	FC	C	H	N	S	O
Petcoke	0.32	10.96	0.17	88.55	89.08	3.64	1.47	5.43	0.20
Orchard coal	1.59	9.27	10.79	78.35	80.80	3.52	1.91	1.14	0.24
Pittsburgh no.8 coal	0.31	35.70	9.80	54.19	76.68	4.93	1.39	1.46	5.42
Illinois#6 coal	3.31	38.46	12.53	45.70	67.50	4.75	1.45	5.72	5.67
Dietz coal	8.46	41.69	3.87	45.99	70.70	4.91	0.58	0.39	11.8
Pust coal	11.12	41.17	8.55	39.16	60.77	4.08	1.28	0.60	13.60
Beulah coal	18.88	39.19	5.19	36.74	54.94	3.87	1.32	0.60	15.20

Note: M, VM, A, FC stand for moisture, volatile matter, ash, fixed carbon respectively.

2.2. Preparation of carbonaceous solid-water slurry. Calculated amount of carbonaceous solids (calculated on the basis of targeted slurry concentration) was added to weighed amount of distilled water. The mixture was then stirred by a propeller agitator for 10 minutes and viscosity was measured [1]. Solids loadings of carbonaceous solid - water slurries were calculated and reported on as received basis.

2.3. Viscosity. Viscosity was measured using Bohlin 88 Viscometer (a product of Malvern Instruments Ltd., U.K.) The shear rate used for all the measurements was $100s^{-1}$ in keeping with the shear rate used in industry. All viscosity measurements were performed in the temperature range of 23 to 26 $^{\circ}$C (ambient temperature).

2.4. Contact Angle. Contact angles were measured using a goniometer (ramé-hart Model 295). Contact angles of water, glycerol and diiodomethane, on the carbonaceous solid surfaces were measured using the Sessile drop technique.

2.5. Zeta Potential. Zeta Potential of the carbonaceous solid samples were measured using Malvern Zetasizer ZS which uses electrophoretic light scattering technique.

3. Results and Discussions

3.1. Interaction Energy Calculations & Analyses. Van der Waal interaction energies and polar interaction energies were calculated on the basis of contact angle measurements whereas the electrostatic interaction energies were calculated using zeta potential measurements. Contact angle values were also used to determine the surface energy parameters. Detailed calculations of the inter-particle interaction energies have been included elsewhere [15]. The observations of the interfacial energy calculation can be summarized as:

1. For each of the concentrated carbonaceous solid–water slurries polar (hydrophobic/hydrophilic) interaction energy is 2 to 3 orders of magnitude greater than the van der Waal or electrostatic interaction energy. Therefore, polar interaction energies play the key role in determining slurry rheology of highly concentrated slurries.

2. For slurries of petcoke, Orchard coal, Illinois no. 6 coal and Dietz coal in water, the inter-particle polar interaction energy is attractive hydrophobic whereas for Pust coal it is repulsive hydrophilic in nature. The magnitude of attractive hydrophobic energy decreases from petcoke to Dietz coal in the order of increasing oxygen to carbon ratio.

Analyses based on the above observations are as follows:

1. Petcoke and Orchard coal particles in water form aggregation network due to the predominant hydrophobic inter-particle interaction energy. The aggregation networks trap free water within the network structure. Since the contact angles of water on petcoke and Orchard coal surface is greater than 100°, the water can only form unstable coalescing droplets on such surfaces. The lack of oxygen functional groups on the surfaces of these solids contribute to its apolarity.[16]

2. The inter-particle hydrophobic interaction energy is less in magnitude for the Illinois no. 6 and the Dietz coal which implies that the aggregation networks in the slurries would be more like loose flocs. Since water contact angle on the surfaces of Illinois no. 6 and Dietz coal are 73° and 42° respectively, water would not form coalescing droplets and actually start forming hydrogen bonds with oxygen functional group on the surface of the carbonaceous solids. Therefore free water from the slurry would start forming stable hydration layers around carbonaceous solids.

3. Formation of aggregation network structures would not be possible in Pust coal water slurries due to the existence of inter-particle hydrophilic repulsive energy. Oxygen functional groups on the surface of such carbonaceous solids form hydrogen bonds with the free water and thus trap free water to form hydration layer around the particles.

4. In all this cases, it is evident that free water is lost from the slurry either in the form of coalescing droplets or hydration layers. Loss of free water reduces fluidity of the slurry and affects the effective solid volume fraction. Solid volume fraction is given by:

$$\text{Solid Volume Fraction } (\varphi) = \frac{Volume\ of\ Solid}{Volume\ of\ Solid + Volume\ of\ free\ water} \tag{1}$$

It is evident from equation (1) that as volume of free water decreases, effective solid volume fraction increases. Now, the volume of free water lost or the volume of trapped water depends on whether the water is lost as coalescing droplets or hydration layers and also on the thickness of hydration layer. Therefore clearly a loss of free water or the increase in effective solid volume fraction, is a function of the surface chemistry of the solid.

3.2. Viscosity Prediction of Carbonaceous Solid-Water Slurries.

The increase in effective solid volume fraction of the carbonaceous solids in suspension with water was determined with the help of measured viscosity values and Krieger-Dougherty equation. The particle-size distributions of the carbonaceous solids were used to calculate maximum solid volume fraction (φ_m) using Veytsman's equation [17]. Since the particle-size distribution was maintained same for all carbonaceous solids-water slurries for this part of the experiment, the calculated φ_m value was 0.74 in all the cases. The experimentally measured viscosities were used in the Krieger-Dougherty equation (given by equation 2) to calculate the final effective solid volume fraction. The factor by which effective solid volume fraction increases or volume fraction correction factor (VFCF) was calculated for all the solids and was found to vary for different carbonaceous solids. VCFC values for each of the carbonaceous solids in water are listed in table 2. For each of these carbonaceous solid-water slurries, initial solids volume fractions were varied till a point until the slurry ceased to flow (i.e. point where the slurry had infinite viscosity).

$$Measured\ Viscosity = \eta_r = \left(1 - \frac{\varphi_f}{\varphi_m}\right)^{-[\eta]\varphi_m} \tag{2}$$

φ_f= final solids loading (dry basis): determined using equation 10

φ_i= initial solids loading (dry basis): already known

It can be observed from table 2 that the VCFC values, increase from hydrophobic to hydrophilic solids implying that the increase in effective solid volume fraction is more for hydrophilic solids. Hydrophilic solids having higher oxygen to carbon ratio can attract more water molecules and form thicker and more stable hydration layers. On the other hand the extremely hydrophobic petcoke and Orchard coal can only trap water in the form of unstable coalescing droplets. It's easier to release trapped water by applying shear in the latter case. A slight variation of VFCF values of a particular carbonaceous solid in water was observed over different solid volume fractions (as indicated by the small standard deviation). The small variation is due to the fact that as the initial solid volume fraction increases, it leads to more and more crowding of the solid particles in the water, leaving less space for the entrapped water or the hydration layer.

Figure 2 shows the variation of VFCF values as a function of the oxygen and carbon content of the carbonaceous solid

Table 2: VFCF values of the different carbonaceous solids in water

	Carbonaceous solids water slurries				
Initial Solids Volume fraction	Petcoke	Orchard coal	Illinois#6 coal	Dietz coal	Pust coal
.36	N/A	N/A	N/A	N/A	1.85
.38	N/A	N/A	N/A	N/A	1.82
.40	N/A	N/A	N/A	1.7	1.77
.42	N/A	N/A	N/A	1.65	1.74
.44	N/A	N/A	1.54	1.59	1.67
.46	N/A	N/A	1.52	1.54	Infinite Viscosity
.48	N/A	1.36	1.47	1.48	
.50	N/A	1.35	1.42	Infinite Viscosity	
.52	1.32	1.33	Infinite Viscosity		
.54	1.31	1.3			
.56	1.27	1.27			
.58	1.24	Infinite Viscosity			
Mean	**1.28**	**1.31**	**1.49**	**1.59**	**1.77**
Standard Deviation	**.03**	**.04**	**.053**	**.07**	**.07**

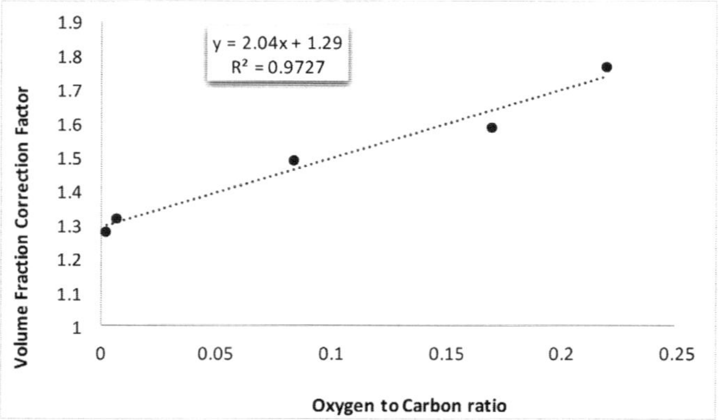

Fig 2: Variation of VFCF with oxygen to carbon ratio

VFCF was observed to vary linearly and correlate well with the oxygen to carbon ratio. Therefore for any carbonaceous solid whose oxygen and carbon content are known , their VFCF can be calculated from equation 3 and used in equation 4 to predict their corresponding slurry viscosities accurately.

$$\boldsymbol{VCFC = 2.04\frac{O}{C} + 1.29} \tag{3}$$

$$\eta_r = \left[1 - \frac{\left(2.04*\frac{O}{C}+1.29\right)\varphi}{\varphi m}\right]^{-[\eta]\varphi m} \tag{4}$$

Where,

O = Oxygen content of the carbonaceous solid on dry basis
C = Carbon content of the carbonaceous solid on dry basis.
Other terms in the equation have already been defined in the previous sections.

3.3. Validation of the modified model.

The modified model was validated using slurries of petcoke, Pittsburgh no.8 coal (bituminous coal) and Beulah coal (lignite coal) particles in water. Maximum packing efficiencies of these samples were determined on the basis of their particle-size distributions using Veytsman's model. Oxygen to carbon ratios as obtained from ultimate analyses of the samples were used in equation 2 to obtain the VFCFs of the corresponding carbonaceous solids in water. Maximum packing efficiencies (φ_m) and VFCFs, inputs to the modified Krieger-Dougherty equation, are reported in table 3.

Table 3: Input parameters to the modified Krieger-Dougherty equation

	Maximum packing efficiency	VCFC
Petcoke	.78	1.28
Pittsburgh no.8 coal	.76	1.43
Beulah coal	.86	1.86

Experimentally measured viscosities of these slurries were then compared with predictions of both Krieger-Dougherty equation and the modified Krieger-Dougherty equation as shown in figures 3 (a-c).

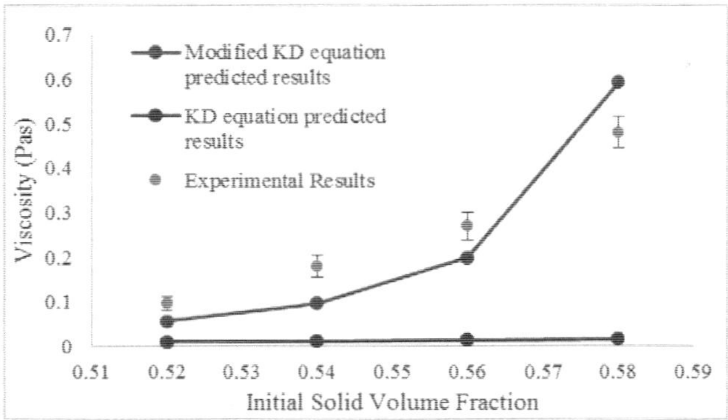

(a) Petcoke-water slurry

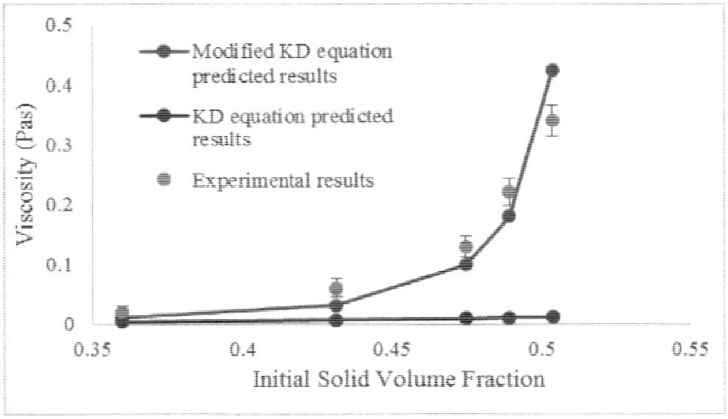

(b) Pittsburgh no.8 coal-water slurry

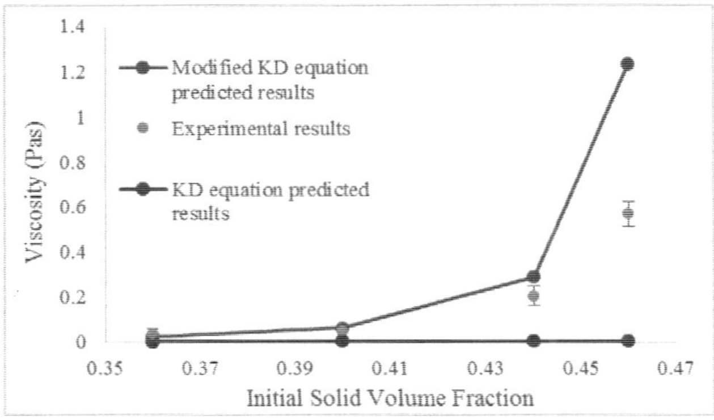

(c) Beulah coal-water slurry

Figure 3: Comparison of experimentally measured viscosities with model predicted viscosities as a function of initial solid volume fraction

It is evident from the figures that modified equation predicts viscosity much more accurately especially at higher solid loading compared to original Krieger-Dougherty equation. At higher solid loading owing to minimal inter-particle distance, inter-particle interaction plays a major role in governing slurry viscosity. The modified equation takes into account the surface interactions, and therefore its predictions are more accurate.

4. Conclusions:

1. Inter-particle polar interaction energies play a dominant role in deciding carbonaceous-solid water slurry microstructure.

2. Carbonaceous solids in water have the tendency to trap free water either in the form of coalescing droplets or in the form hydration layers. The loss of a part of the free water from the slurries reduces the fluidity of the slurry by increasing the effective solid volume fraction. The increase in solid volume fraction is characteristic of a particular carbonaceous solid in water and therefore represents surface chemistry of the carbonaceous solid.

3. Increase in solid volume fractions of carbonaceous solids in water were calculated using measured viscosity values and Krieger-Dougherty equation. The increase in solid volume fraction was found to increase with increase in oxygen functional groups present on the surface of the solids

4. VCFC (increase in effective solid volume fraction) was observed to correlate well with the oxygen to carbon ratio of the carbonaceous solid. Therefore for a carbonaceous solid with known oxygen to carbon ratio, its VCFC can be predicted from the correlation and by using the VCFC value in the modified Krieger-Dougherty equation its viscosity can be predicted accurately.

5. Modified model was validated with three other sample slurries and was found to make accurate prediction of slurry viscosity.

5. Future Work:

- Volume of loss of free water for each of carbonaceous solid water slurries is to be confirmed by Differential Scanning Calorimetry.
- This model would be further developed to take into account effect of additives and slurry pH on slurry viscosity.
- Expand the applicability of the model so as to predict viscosities of even non-carbonaceous solid-water slurries such as proppant water slurry.

6. References:

[1] N.-S. Roh, D.-H. Shin, D.-C. Kim, and J.-D. Kim, "Rheological behaviour of coal-water mixtures. 1. Effects of coal type, loading and particle size," *Fuel*, vol. 74, no. 8, pp. 1220–1225, Aug. 1995.

[2] F. Gao, J. Liu, C. Wang, J. Zhou, and K. Cen, "Effects of the physical and chemical properties of petroleum coke on its slurryability," *Petroleum Science*, vol. 9, no. 2, pp. 251–256, Jul. 2012.

[3] H. Dinçer, F. Boylu, a. . Sirkeci, and G. Ateşok, "The effect of chemicals on the viscosity and stability of coal water slurries," *International Journal of Mineral Processing*, vol. 70, no. 1–4, pp. 41–51, Jun. 2003.

[4] S. Mueller, E. W. Llewellin, and H. M. Mader, "The rheology of suspensions of solid particles," *Proceedings of the Royal Society of London A: Mathematical, Physical and Engineering Sciences*, vol. 466, no. 2116, pp. 1201–1228, Feb. 2010.

[5] M. Mooney, "The viscosity of a concentrated suspension of spherical particles," *Journal of Colloid Science*, vol. 6, no. 2, pp. 162–170, Apr. 1951.

[6] R. Simha, "A Treatment of the Viscosity of Concentrated Suspensions," *Journal of Applied Physics*, vol. 23, no. 9, p. 1020, 1952.

[7] A. Mukherjee, P. Rozelle, and S. V Pisupati, "Influence of Surface Properties and Applicability of Current Viscosity Models for Carbonaceous Solid Water Slurries," accepted.

[8] T. Tadros, "Interparticle interactions in concentrated suspensions and their bulk (rheological) properties.," *Advances in colloid and interface science*, vol. 168, no. 1–2, pp. 263–77, Oct. 2011.

[9] I. M. Krieger, "A Mechanism for Non-Newtonian Flow in Suspensions of Rigid Spheres," *Journal of Rheology*, vol. 3, no. 1, p. 137, 1959.

[10] W. Pabst, "Fundamental Considerations On Suspension Rheology," *Ceramics-Silikaty*, vol. 48, pp. 6–13, 2004.

[11] J. P. Dooher, N. Malicki, and J. Trudden, *Physiochemical principles of coal-water-slurry gasifier feedstocks*. 1990, p. 151.

[12] H. Usui, "A thixotropy model for coal-water mixtures," vol. 60, 1995.

[13] H. Usui, "Viscosity Prediction of Dense Slurries with Agglomerative Coal Particles.," *Nihon Reoroji Gakkaishi*, vol. 29, no. 2, pp. 71–76, 2001.

[14] M. Pawlik, J. S. Laskowski, and H. Liu, "Effect of Humic Acids and Coal Surface Properties on Rheology of Coal-Water Slurries," *Coal Preparation*, vol. 18, no. 3–4, pp. 129–149, Jan. 1997.

[15] A. Mukherjee and S. V Pisupati, "Inter-particle interactions in highly concentrated coal-water slurries and their effect on slurry viscosity," *Energy & Fuels*, submitted.

[16] G. D. Botsaris, *Interfacial Phenomena in Coal Technology*. 1988, p. 464.

[17] B. Veytsman, J. Morrison, A. Scaroni, and P. Painter, "Packing and Viscosity of Concentrated Polydisperse," *Energy & Fuels*, vol. 0624, no. 6, pp. 1031–1039, 1998.

401697 Optimizing Crude Oil Separations – a Study of Turbulence and Chemical Composition Influence on the Drop Size Distribution of Crude Oil in Maritime Conditions (Electronic Poster Presentation)

Abstract:

Optimizing crude oil separations – a study of turbulence and chemical composition influence on the drop size distribution of crude oil in maritime conditions

Efficiency in the exploration of crude oil deposits is of great importance for the oil and gas industry. With the usage of large amounts of feed water in the process comes the need for fast and efficient separation processes for the liquid/liquid dispersion. The quality of the separation translates directly into costs for storage, transportation and subsequent processing. The dispersion is thermodynamically unstable and given time, the droplets will separate out to form two bulk liquid phases. The drop size distribution is the most relevant parameter for this type of separation process. Great efforts have been made to measure the size of liquid dispersed droplets. The utilized techniques can be divided into those that require samples to be withdrawn from the vessel or pipe, and those that can be applied in situ.

For deep sea analysis in-situ methods are preferential since they account of the significant impact of the present conditions on the size of the drops. Many investigations of droplet size distributions or a representable diameter like d_{32} have focused on highly dilute dispersions, where it is assumed that coalescence can be eliminated. As this approach provides results with limited practical application, a measurement technology for high concentrations is necessary. In comparison to chemical, pharmaceutical, and food manufacturing, the petroleum industry presents additional challenges, as the optical conditions of oil streams are much more complex then specialty and fine chemicals. Also the variation of fluid physical properties and process conditions is stronger. The understanding, prediction and control of the drop size in crude oil applications is rather challenging. A detailed description of the opposing phenomena, drop breakage and coalescence, is paramount for successful mathematical modeling. An additional difficulty in handling such dispersed systems is the probability of phase inversions (PI). In case of a spontaneous occurrence within a process, the consequences for product and equipment can be catastrophic [1]. On the contrary, it can be used for a more controlled handling of the described applications. Therefore a better understanding and modeling of this phenomenon is desirable.

This study reports on a turbulent liquid/liquid system using original North Sea crude oils and sea water was investigated. The operation parameters dispersed phase fraction, turbulence intensity and surfactant concentration have been varied systematically. The transient drop sizes as the unstable behavior of the water and organic phase have been observed using an in-situ microscope [2]. The drop size was measured using a fully automated image analysis tool [3]. These data give a reliable base to understand liquid-liquid separations better.

[1] Nienow, A. W., Adv. Colloid Interface Sci. 2004, 108, 95-103.

[2] Maaß, S., Wollny, S., Voigt, A., Kraume, M., Exp. Fluids 2011, 50, 259-269.

[3] Maaß, S., Rojahn, J., Hänsch, R., Kraume, M., Comput. Chem. Eng. 2012, 45: 27-37.

401743 Prediction of High-Temperature and High-Pressure Dynamic Viscosity of Hydrocarbons By Van Der Waals Transport Equation of State

Abstract:

According to the review of liquid viscosity by Brush (1962), [1] "there is no general agreement on whether viscosity is essentially due to the attractive and repulsive forces," but on macroscopic scale both viscosity and density are function of the thermodynamic equilibrium state of the fluid[2] (therefore, they both surrender to the Law of Corresponding States [3,4]) while on microscopic scale both properties reflect the effects of molecular motion and interaction. So, in accordance with the hypothesis enunciated by Phillips (1927)[5] and later improvements by Little-Kennedy (1966)[6] and Lawal (1986), [7] a method is established for predicting Newtonian viscosity or dynamic viscosity (η) over the entire PVT states of single-phase fluids such as polar and nonpolar substances, hydrocarbon, non-hydrocarbon, petroleum fractions and high-molecular-weight-fluids of the type currently being measured for high-temperature and high-pressure (HTHP) deep-water formations [8-17] and carbon dioxide sequestration.

By using the1986 formalism of the Viscosity Equation of State (VEOS) development, the VEOS technique is designed on the basis of the phenomenological similarity between PVT and TηP spinodal graphs and is solely depended on accurate critical viscosity of pure substances. There is no resorting to the judicious choice of liquid density correlation or any volumetric properties because the coexistence gas-liquid viscosities, the dilute-gas viscosity, the dense and supercritical fluid viscosities are based on the four-parameter cubic equation of state[7] which is valid over the entire PVT states, include the vapor-liquid critical point.

The VEOS is validated by the prediction of accurate coexistence gas-liquid viscosities over the entire PVT states of pure substances, including polar and nonpolar substances, inert gases (Argon, Helium, Xenon), hydrocarbons of high-molecular-weights and non-hydrocarbons (nitrogen, carbon dioxide, hydrogen sulfide) and liquid viscosity of very viscous fluids.

References:

1. Brush, S. G., "Theories of Liquid Viscosity," *Chem. Rev. 62*, 513, **1962**
2. Brule, M. R., Starling, K. E., *Ind. Eng. Chem. Process Des. Dev.*, *23*, 833, **1984**
3. Ely, J. F., Hanley, H. J. M., "Prediction of Transport Properties: 1. Viscosity of Fluids and Mixtures," *Ind. Eng. Chem. Fund. 20*, 323-332, **1981**
4. Ely, J. F., Hanley, H. J. M., "Prediction of Transport Properties: 2. Thermal Conductivity of Pure Fluids and Mixtures," *Ind. Eng. Chem. Fund. 22*, 90-97, **1983**
5. Phillips, P., "The Viscosity of Carbon Dioxide," *Proc. Roy. Soc. 87A*, 48, **1927**
6. Little, J. E., Kennedy, H. T., "A Correlation of the Viscosity of Hydrocarbon Systems with Pressure, Temperature and Composition," *Soc. Pet. Eng. J.*, 157 (June**1968**)
7. Lawal, A. S., "Prediction of Vapor and Liquid Viscosities From the Lawal-Lake-Silberberg Equation of State," In *SPE/DOE Paper No.14926 in the Proceeding of SPE/DOE Fifth Symposium on Enhanced Oil Recovery of the Society of Petroleum Engineers and the Department of Energy* held in Tulsa, OK, April 20-23, **1986**
8. Caudwell, D. R.; Vesovic, V.; Trusler, J. P. M.; Wakeham, W. A., "The viscosity and Density of n-Dodecane and n-Octadecane at Pressures up to 200 MPa and Temperatures up to 473K," *Int. J. Thermophysics*, *25*, 1340-52, **2004**
9. Caudwell, D. R.; Trusler, J. P. M.; Vesovic, V.; Wakeham, W. A., "Viscosity and Density of Five Hydrocarbon Liquids at Pressures up to 200 MPa and Temperatures up to 473 K," *J. Chem. Eng. Data*, **54**, 359-366, **2009**
10. Y.L. Sen, Kiran, E., "High-pressure viscosity and density of n-alkanes," *Int. J. of Thermophysics, 13,* 411-442, **1992**
11. Burgess, W. A., Tapriyal, D., Gamwo, I. K., Morreale, B. D., McHugh, M. A., Enick, R. M., "Viscosity Models Based on the Free Volume and Frictional Theories for Systems at Pressures to 276 MPa and Temperatures to 533 K," *Ind. Eng. Chem. Res.*, *51*, 16721-33, **2012**
12. Liu, K.; Wu, Y.; McHugh, M. A.; Baled, H.; Enick, R. M.; Morreale, B. D., "Equation of State Modeling of High-Pressure, High-Temperature Hydrocarbon Density Data," *J. Supercrit. Fluids*, *55*, 701, **2010**
13. Baled, H. O., Xing, D., Katz, H., Tapriyal, D., Gamwoa, I. K., Soong, Y., Bamgbade, B. A., Wu, Y., Liu, K., McHugh, M. A., Enick, R. M., "Viscosity of n-Hexadecane, n-Octadecane and n-Eicosane at pressures up to 243 MPa and temperatures up to 534 K," *J. Chem. Thermodynamics*, *72*, 108-116, **2014**
14. Ducoulombier, D.; Zhou, H.; Boned, C.; Peyrelasse, J.; Saint-Guirons, H.; Xans, P., "Pressure (1−1000 bar) and Temperature (20−100 °C) Dependence of the Viscosity of Liquid Hydrocarbons," *J. Phys. Chem.*, *90*, 1692-1700, **1986**
15. Hogenboom, D. L.; Webb, W.; Dixon, J. A., "Viscosity of Several Liquid Hydrocarbons as a Function of Temperature, Pressure, and Free Volume," *J. Chem. Phys.*, *46*, 2586-98, **1967**
16. Tanaka, Y.; Hosokawa, H.; Kubota, H.; Makita, T., "Viscosity and Density of Binary Mixtures of Cyclohexane with n-Octane, n-Dodecane, and n-Hexadecane Under High Pressures," *Int. J. Thermophysics*, *12*, 245-264, **1991**

17. Assael, M. J.; Papadaki, M., "Measurements of the Viscosity of n-Heptane, n-Nonane, and n-Undecane at Pressures up to 70 MPa," *Int. J. Thermophysics*, 12, 801–810, **1991**

401790 Inversion of the Van Der Waals Parameters for Estimation of Binary Interaction Constants

Abstract:

As yet, it is still impossible to invert the equations of statistical mechanics to explicitly yield intermolecular forces as functions of the macroscopic properties of individual pure substances. But, the critical point of the van der Waals (VDW) gas-liquid transition provides a source of deriving interaction constants from the measured critical temperatures and pressures of binary systems. By inverting the three-dimensionless criticality parameters established by the VDW theory, we back-calculate the binary interaction constants for the attractive parameter and two other EOS parameters whose binary interaction constants relate mixture critical volume and compressibility factor of VDW equation to composition. The Lawal-Lake-Silberberg (LLS) equation of state (EOS) provides a prototype procedure for VDW type-equations in establishing binary interaction constants from measured critical loci of binary systems and the technique is validated by the prediction of accurate critical properties, high-pressure volumetric properties and binary high-pressure phase equilibria. It is revealed that the derived binary interaction constants are *totally* independent of temperature, pressure and composition. However, the generalization of the binary interaction constants in terms of the macroscopic properties of pure substances is still very challenging but may not be impossible through the realms of statistical mechanics.

401284 Polymer Solution Preheating to Achieve Steam Usage Reduction in Polymer Isolation By Precipitation Jets

Abstract:

Isolation of polymers from its solution (e.g. hydrocarbon solvents) can be achieved by direct contact with saturated steam, typically in a precipitation jet. Polymer precipitates out of the solution due to solvent evaporation by steam. During this heat exchange, steam condensation leads to a wet polymer powder. This moisture in the solid polymer needs to be removed in the subsequent drying operation to meet the finished product specifications for extrusion performance. The driving force for separating the moisture from the polymer particles is the concentration gradient with the drying medium and is controlled by the outward diffusion of the water from the powder. This makes the drying operation highly energy intensive and accounts for a major fraction of the manufacturing costs. Hence, a strong interest exists in improving the efficiency of dryers for reducing costs and carbon footprint.

In this work, we demonstrate a significant reduction in steam usage by preheating the polymer solution before feeding it into the precipitation jets. The preheating of the solution causes reduced moisture in the rein powder obtained in the precipitation jet. This moisture reduction in the polymer powder exiting the precipitation jets leads to lowering in energy requirement for powder drying in the post-drying stages. In an overall energy balance of the polymer isolation operation, we have demonstrated at least 10% reduction in energy usage through steam savings and consequently reduced the carbon footprint. Furthermore, we are able to improve the production efficiency due to the reduction in the requirement of steam/polymer weight ratio.

15 AIChE
Spring Meeting, Austin, TX

Impact of High-Emissivity Coatings on Process Furnace Heat Transfer

Paper 85b

Bradley Adams
Reaction Engineering International
746 E. Winchester St., Suite 120
Murray, UT 84107
adams@reaction-eng.com

Martin Denison
Reaction Engineering International
746 E. Winchester St., Suite 120
Murray, UT 84107
denison@reaction-eng.com

John Olver
Emisshield, Inc.
2000 Kraft Drive, Ste 2600
Blacksburg, VA 24060
john.olver@emisshield.com

Prepared for Presentation at
American Institute of Chemical Engineers
2015 Spring Meeting
11th Global Congress on Process Safety
Austin, Texas
April 27-30, 2015

Impact of High-Emissivity Coatings on Process Furnace Heat Transfer

Paper 85b

Bradley Adams
Reaction Engineering International
746 E. Winchester St., Suite 120
Murray, UT 84107
adams@reaction-eng.com

Martin Denison
Reaction Engineering International

John Olver
Emisshield, Inc.

Keywords: high emissivity, coating, enhanced heat transfer, radiant efficiency, process furnace

Abstract
High emissivity coatings have been shown to enhance heat transfer in high temperature process furnaces. This paper examines the theory behind the enhanced heat transfer, approaches for simulating the effects of high emissivity coatings, simulation results evaluating the impact of high emissivity coating in an idealized pyrolysis furnace, field results from an industrial furnace where one such coating from Emisshield, Inc. was installed, and a summary of reported field results for larger chemical process furnaces.

High emissivity coatings on furnace walls absorb incident radiant energy and re-emit this energy across the wavelength spectrum, including emission through transparent windows in the spectrum. This spectral redistribution of emitted energy allows more radiant energy to pass through the combustion flue gas and be transferred to the process surface. Simulation of this process requires spectral based radiation models which can represent the spectral absorption bands and transparent windows in the flue gas. This is accomplished using a Spectral-Line-Weighted (SLW) model. CFD simulation of a pyrolysis furnace showed that changing the refractory emissivity from 0.4 to 0.9 resulted in negligible differences when using only a grey gas model. When the same change was simulated using a spectral gas property model, the radiant furnace efficiency increased by 1.5% and the arch temperature decreased by 29°C. Simulating this behavior is helpful when evaluating adjustments to furnace operation, for example when changing from a liquid to gas process feedstock.

Data from successful applications of a high emissivity coating technology showed fuel savings of 7-8% in a glass melting furnace and 5-6% in cracking furnaces due to more efficient heat transfer. Production output and run lengths between decoke cycles were also reportedly enhanced in several cracking furnace applications.

1. Introduction
High temperature process furnaces are used in a variety of industrial processes. A key to process efficiency and throughput is efficient heat transfer from flames to the process material or process

tubes. One technology that has been shown to enhance heat transfer in high temperature process furnaces is high emissivity coatings. These coatings work as follows.

For gas-fired furnaces, normal furnace refractories tend to reflect a majority of arriving energy back into the furnace flue gas at the same spectral wavelength at which it is emitted from the flue gas. This energy is then again absorbed by the flue gas, limiting the amount of energy transferred to the process (tubes or materials). High emissivity coatings on furnace walls absorb more of the incident radiant energy and re-emit this energy across the wavelength spectrum, including emission through transparent windows in the spectrum. This spectral redistribution of emitted energy allows more radiant energy to pass through the flue gas and be transferred to the process surface.

Figure 1 illustrates the difference in radiant emissive power between the full blackbody spectrum and the absorption/emission bands in combustion gases. Radiation that is emitted in the combustion gases bands, and then reflected from a low emissivity surface back into the gases, will be reabsorbed by the gases. Radiation that can be absorbed by a high emissivity surface and re-emitted in a manner that more closely resembles the full blackbody spectrum will more efficiently transmit radiation between the gas absorption bands.

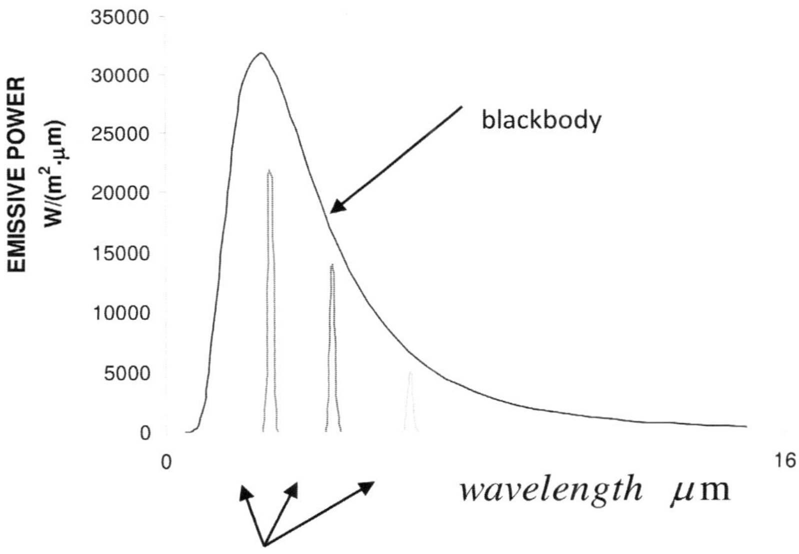

Figure 1. Emissive power spectrum showing blackbody behavior compared to absorption/emission bands from combustion gases.

Figure 2 illustrates the dependence of emissive power on different surface emissivities at a specified temperature. The blackbody surface is represented by an emissivity of 1.0. This is the most efficient emitter. A surface with an emissivity of 0.9 (say, a surface coated with a high emissivity coating) provides a relatively good approximation of the blackbody emission. A more typical surface emissivity of 0.4 from standard refractory or batting materials is a much less efficient emitter. Thus a typical surface not only reflects much of the arriving radiant energy back to the gases at wavelengths subject to absorption, but also emits what energy is absorbed on the surface less efficiently.

High emissivity technology developed by NASA to protect the space shuttle during re-entry was licensed to Emisshield, Inc., which developed a family of high emissivity coatings having service temperatures of 1700°C and above. EMISSHIELD® is a registered trademark for coatings manufactured by Emisshield, Inc., and is covered by U.S. Patent 6,921,431. Figure 3 illustrates surface emissivity as a function of temperature with and without the Emisshield high emissivity surface coating. The coating emissivity remains nearly constant with changed wall temperature, whereas the emissivity of the uncoated refractory surface decreases with increasing temperature.

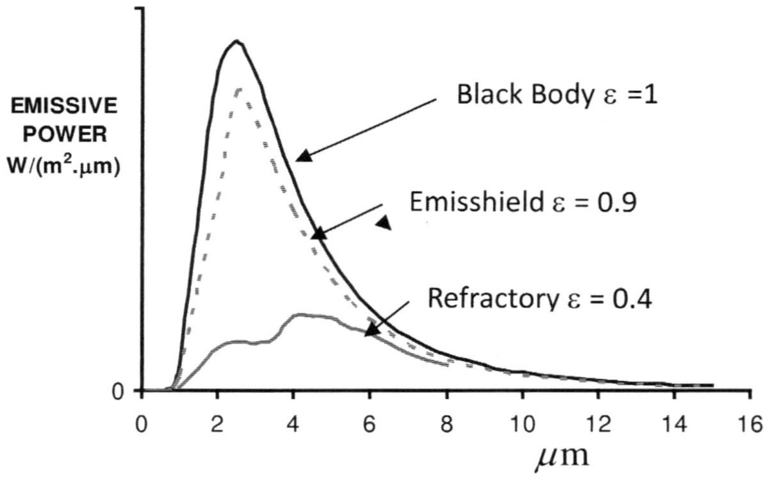

Figure 2. Emissive power spectrum comparing blackbody behavior to surfaces with lower emissivities.

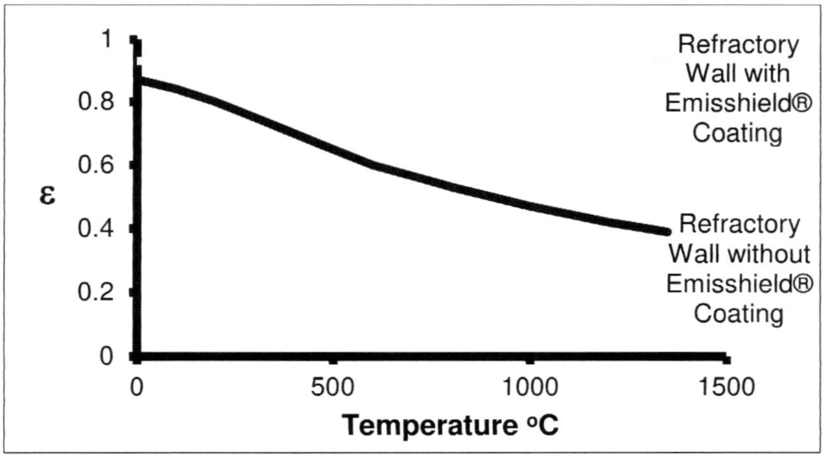

Figure 3. Refractory surface emissivity as a function of temperature with and without the Emisshield surface coating.

2. Simulation Approach

Assessment of the effectiveness of the high-emissivity coating requires a modeling approach capable of replicating the spectral radiation effects described above. REI's *ADAPT* CFD code was used in this study. The characteristics of this software have been previously described [1, 2]. *ADAPT* capabilities include:

- localized mesh refinement to capture burner geometry details and near-burner mixing;
- discrete ordinates method with SLW gas property model for solving radiative transfer equation;
- reduced chemical kinetic mechanisms with *in-situ* adaptive tabulation (ISAT) to compute finite rate chemistry;
- a combination of conventional Eulerian turbulence modeling with Monte-Carlo PDF methods to model turbulent reactions and mixing; and
- a matrix-free Newton-Krylov method to reduce solver computational time and improve robustness.

Radiation property models can and usually have a close relationship to the radiative transfer equation (RTE) solution method. For example, efficient but accurate predictions of gas radiative transfer can be a challenge due to the strong spectral variation of the gas radiative properties (absorption coefficient). The absorption spectra of gases consist of many thousands of spectral lines because gases emit and absorb electromagnetic radiation only at discrete frequencies where the corresponding photon energies match the quantum changes in energy of the gas molecules. Narrow- and wide-band models for gas radiation properties provide a representation of spectrally averaged quantities such as transmissivity or band absorptance that require specification of path-length. Therefore, these band models do not lend themselves to arbitrary solution methods of the RTE, which, in its most fundamental form, is written in terms of a monochromatic absorption coefficient at a specific wave number and not in terms of spectrally averaged properties. The discrete ordinates model [3] for solving the RTE was used in this study and also requires the gas radiative properties to be specified as an absorption coefficient.

Gas radiation models have been developed over the past few decades, where the main idea of the models is replacement of integration over wave number with integration over absorption cross-section with the reordering of continuous gas absorption coefficient into a set of gray gases. The total gas emissivity is expressed in the weighted-sum-of-gray gases form as

$$\epsilon = \sum_j a_j (1 - e^{-\kappa_j L}) \qquad \text{[Eq. 1]}$$

where a_j are the black body weights and κ_j are the associated absorption coefficients. The RTE in expressed in the weighted-sum-of-gray gas form is given as

$$\frac{dI_j}{ds} = \kappa_j \left(a_j \frac{\sigma T^4}{\pi} - I_j\right) \qquad \text{[Eq. 2]}$$

To obtain spectrally integrated quantities one simply sums over the index j. The total incident radiative heat flux, for example, is given as

$$q_{inc} = \sum_j q_{inc,j} \qquad \text{[Eq. 3]}$$

where the incident flux, $q_{inc,j}$, is obtained from the solution of the RTE for that index.

The spectral line-based weighted-sum-of-gray-gases (SLW) model, first introduced by Denison and Webb [4], was developed based on detailed line-by-line spectral data of gases. The basis of this model is the use of the Absorption Line Blackbody Distribution Function (ALBDF) [5]. Denison and Fiveland [6] later extended the spectral based weighted-sum-of-gray gases model to Edward's wide band model [7] parameters via a wide-band black-body distribution function. In this study, the correlation of Denison and Fiveland [6] was used but with only one gray gas and one clear gas. The clear gas represents the spectral windows between the gas bands where the gas is not participating radiatively. Denison and Fiveland's correlation was used to calculate black body weights of the WSGG, while the corresponding gray gas absorption coefficients were determined to match the total emissivity based on a large number of gray gases.

3. Sample Pyrolysis Furnace Case

An idealized pyrolysis furnace was modeled with *ADAPT* to examine the impact of using a high-emissivity coating on the furnace walls. One-half of the symmetric furnace geometry was modeled as shown in Figure 4. The half-furnace was fired at 15.26 MW, with a fuel comprised of 50% methane, 50% hydrogen. The furnace was first simulated with the furnace walls at an emissivity of 0.4 (Case 1), then again with an emissivity of 0.9 (Case 2). Process coils were assigned an emissivity of 0.85 for both cases. All emissivities were assumed to be constant over the spectrum of wavelengths for these analyses. Spectral simulations (as described above) were used to calculate furnace properties for the two cases. Note that if spectral calculations were not used, the heat transfer and temperature results for the two cases would be identical.

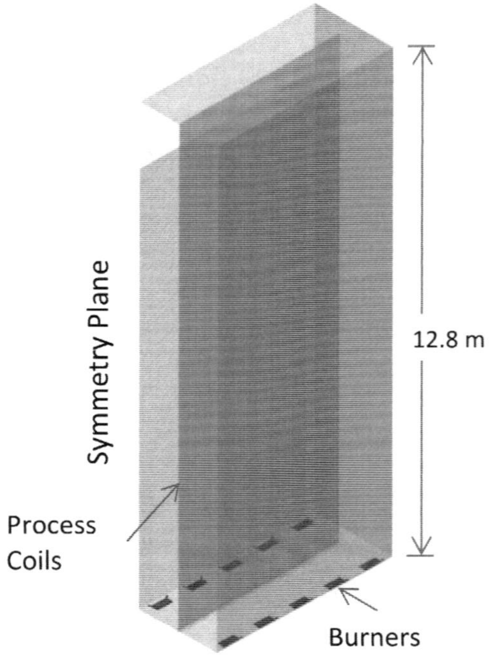

Figure 4. Pyrolysis furnace geometry.

Table 1 summarizes the results for Case 1 and Case 2 scenarios. Only the wall emissivities were varied between the two cases. The higher wall emissivity for Case 2 resulted in more of the flame energy being transferred to the process tubes. As a result, the process fluid outlet temperature rose by 4 °C, the maximum tube surface temperature rose by 4 °C and the furnace thermal efficiency increased from 39.0% to 40.5%. Because the casing loss was similar for both cases, the higher energy transfer to the process tubes caused the energy remaining in the flue gas to decrease and the flue gas exit temperature dropped by 29°C. These differences can be significant with respect to furnace process yield and/or required furnace firing rates.

Table 1. Pyrolysis Furnace Simulation Results.

	Case1	Case2
Wall Emissivity	0.4	0.9
Firing Rate (MW)	15.26	15.26
Flue Gas Exit Temperature (°C)	1,223	1,194
Tube Heat Absorption (MW)	5.95	6.18
Max Tube Temperature (°C)	1,352	1,356
Process Outlet Temperature (°C)	855	859
Casing Loss (% of firing rate)	1.6	1.6
Furnace Radiant Efficiency (%)	39.0	40.5

Figure 5 shows the flame shapes in the furnace as defined by a 5000 ppmv CO iso-surface. The iso-surfaces are colored by flame temperature. The flames were longer than typical flames from floor burners due to the idealized nature of the burners. Flame temperatures peaked approximately half-way up the furnace. Flame shapes for Case 1 and Case 2 were indistinguishable, suggesting that the changes to heat transfer and flame temperature did not affect the flame geometry.

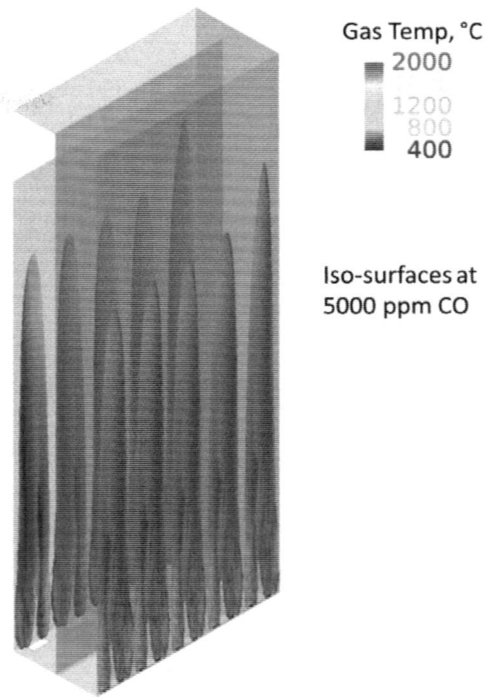

Figure 5. Flame shapes as defined by 5000 ppm CO iso-surface.

Figure 6 shows the gas temperature profiles inside the furnace for Case 1 and Case 2. As more energy is removed from the flame or flue gas and transferred into the process tubes, the flue gas temperature drops. Consistent with the 29°C drop in flue gas exit temperature for Case 2, the furnace gas temperature profile is slightly lower for Case 2. The difference is most pronounced at the top of the furnace where longer flue gas residence times have allowed more heat transfer.

Figure 7 plots the wall surface temperature for Case 1 and Case 2. The profiles and magnitudes are very similar, but Case 2 has slightly higher wall temperatures, consistent with the higher energy absorption and re-emission at the higher emissivity. Similarly, the Case 2 process tubes also have a slightly higher tube temperature, which in turn produces higher heat transfer to the process fluid.

To more clearly illustrate the differences in radiative heat flux between the cases, the incident heat flux at the process tubes (averaged over all tubes) is plotted as a function of furnace height in Figure 8. The higher emissivity case has a higher incident flux over the entire furnace height. Measurements of heat absorption in large process furnaces suggest that the flux differences may be greater than shown in the model results. Reasons for potential discrepancies include differences in flame temperature, flue gas composition, surface temperature, process tube temperature, and volume to surface ratios between actual furnaces and the idealized furnace box modeled here.

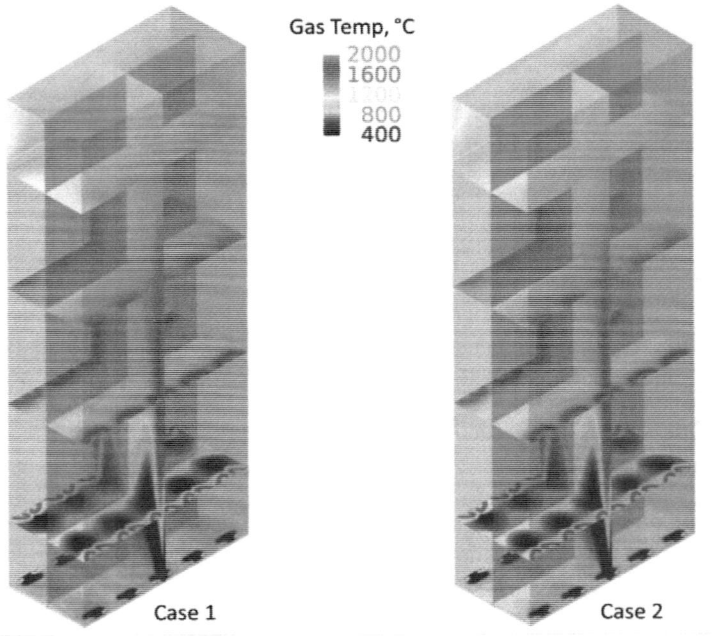

Figure 6. Gas temperature profiles for Case 1 and Case 2.

These CFD results do illustrate that the higher emissivity coating does improve the transfer of heat from the flame to process tubes, by making the radiative heat transfer between the furnace walls (as heated by the flame) and the process tubes more efficient. The overall effect is higher furnace efficiency. Simulating this behavior is helpful when evaluating adjustments to furnace operation, for example when changing from a liquid to gas process feedstock.

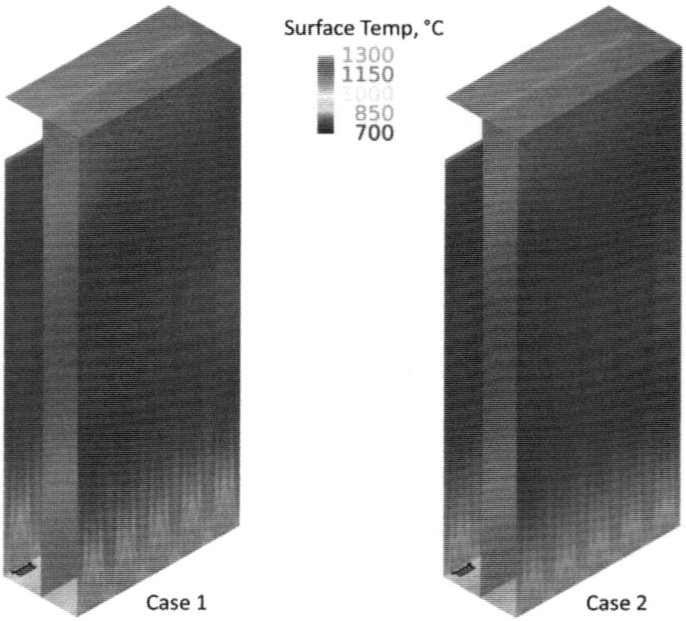

Figure 7. Furnace wall and process tube surface temperatures for Case 1 and Case 2.

This sample analysis did not consider temperature or wavelength-dependent spectral characteristics of the furnace wall and tube emissivities (i.e., variations in wall and tube emissivity based on temperature and wavelength). Including these details may change the magnitude of the predicted heat fluxes, but is not expected to change the conclusion that high emissivity refractory walls produce enhanced heat transfer to the process tubes. The analysis also did not account for any surface conductivity changes associated with application of a coating that could more evenly distribute the absorbed heat before re-emission.

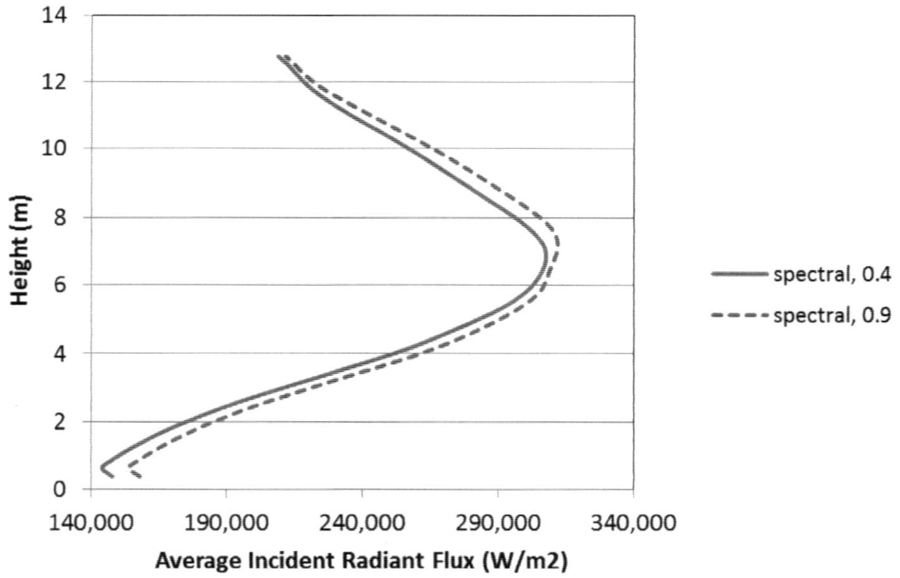

Figure 8. Average incident heat flux at the process tubes for Case 1 (ε_w=0.4) and Case 2 (ε_w=0.9).

4. High Emissivity Coating Field Applications

4.1 Glass Furnace

A glass melting furnace is used to melt a batch of recycled glass (cullet) by heat transfer from gas burners and hot refractory walls to the cullet and melt. The use of high emissivity coatings on the furnace walls can potentially increase the efficiency of heat transfer to the cullet and melt, resulting in reduced firing rates to achieve a molten state for the glass. The concept for this application is shown in Figure 9. When high emissivity coatings are applied to the hot face of the refractories in the furnace superstructure and crown, radiant and convective energy from the burners and hot furnace gases are absorbed at the surface of the coating and re-radiated to the cooler glass batch. The use of the coating increases the heat transfer from the flame and furnace walls to the melt. This results in lower wall temperatures, lower gas temperatures (including less energy loss via flue gas), and higher melt temperatures. Alternately, the same melt temperature is achievable with lower firing rates (i.e., lower wall and gas temperatures).

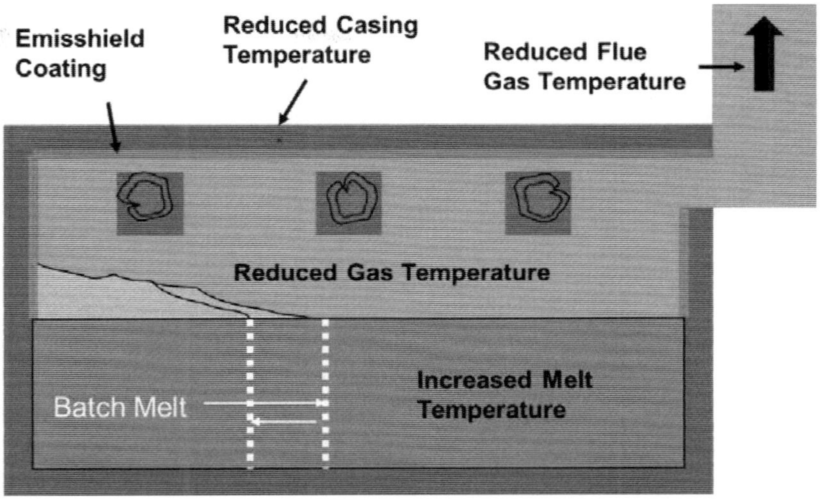

Figure 9. Glass furnace schematic illustrating temperature impacts of coating.

An Emisshield high-emissivity coating was applied to the crown and superstructure in an Owens Corning glass furnace. This application has been previously documented [8] and is only summarized here to illustrate the impact of the coating on furnace heat transfer.

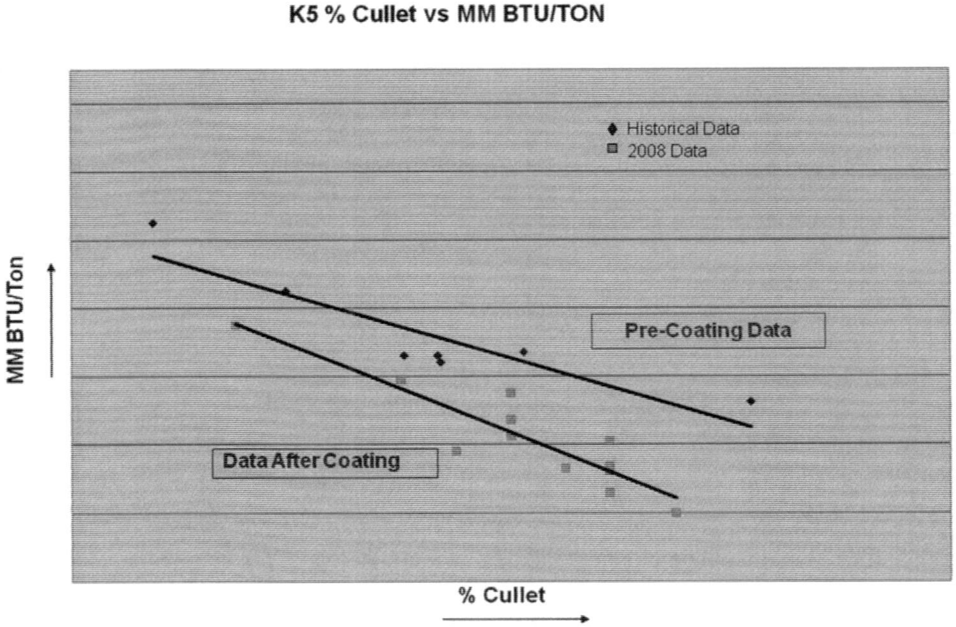

Figure 10. Glass furnace energy requirements with and without coating.

The impact of the coating is shown in Figure 10. The top curve of Figure 10 shows the energy consumption in the furnace at various cullet levels over the previous campaign. This plot confirms the well-established relationship that as cullet level in the batch is increased the energy consumption will decrease. These data points represent many months of operation at each level and thus are considered reliable for comparison to the energy consumption after installation of the high emissivity coating. The bottom curve is the energy consumption after the coating was applied to the furnace. Cullet levels in the batch were changed to be able to compare directly to the pre-coating energy consumption. Care was taken to spend enough time at each cullet level to be sure any change in energy consumption was real. Clearly the energy consumption is less after application of the coating. Although the axes are not labeled for proprietary reasons, the energy saving was in the range of 7-8%.

According to the glass furnace owner, the coating was still providing energy savings after two and one-half years of operation. Since the application and data collection on this furnace, the owner has applied a high emissivity coating to other oxy-fuel furnaces and to one gas-fired furnace. Furnaces identical to the one discussed here have shown a similar 7-8% energy savings while other oxy-fuel furnaces have shown savings as high as 10%. Overall, the Emisshield high-emissivity coating has been applied to more than 80 glass furnaces worldwide over the past seven years.

4.2 Cracking Furnaces

Table 2 lists furnace owner-reported performance results for application of the Emisshield high emissivity coating in several ethylene and EDC cracking furnaces. The trends are similar to the glass furnace discussed above. Application of the high emissivity coating to the furnace refractory resulted in enhanced heat transfer to the process tubes. This allowed plant operators to reduce firing rate (and fuel consumption) and still maintain sufficient, or some cases increased, production. An added benefit appeared to be a reduction in coking rates, i.e., increased time between decoking, in the process tubes. Coking rate is typically associated with hot spots or high heat flux on the process tubes. A reduced coking rate suggests that the local heat flux and/or tube temperatures were less with the coating, likely as a result of the reduced firing rates attainable with improved heat transfer efficiency. Emisshield's high emissivity coating has also been applied to several VCM furnaces in the past five years, but performance data for these systems has not been reported yet.

Table 2. Selected Chemical Process Furnace Applications with Emisshield High Emissivity Coatings.

Location	Furnace Type	Appl. Date	Initial Results Reported by Owners
Port Arthur, TX	Ethylene Cracking Furnace #2	JUN 2005	12% Fuel Savings Reported
Longview, TX	Ethylene Cracking Furnace #2	APR 2005	6% Fuel Savings; 9% Steam Reduction
Longview, TX	Ethylene Cracking Furnace #3	MAY 2006	6% Fuel Savings; 9% Steam Reduction
Longview, TX	Ethylene Cracking Furnace #1	JUL 2007	6% Fuel Savings; 9% Steam Reduction
Baton Rouge, LA	EDC Cracking Furnace #1	NOV 2007	5% Fuel Savings; 6% Production Output More than doubled the time between decoking
Baton Rouge, LA	EDC Cracking Furnace #3	SEP 2008	5% Fuel Savings; 6% Production Output More than doubled the time between decoking
Baton Rouge, LA	EDC Cracking Furnace #2	NOV 2008	5% Fuel Savings; 6% Production Output More than doubled the time between decoking
Mideast	VCM	2013/2014	10-15% increase in Production

4. Conclusions

The concept of how high emissivity coatings can enhance heat transfer efficiency in process furnaces has been reviewed. The enhanced efficiency can lead to increased process production and/or reduced firing rate (fuel) requirements in combustion-heated process furnaces. Modeling the behavior of high emissivity coatings can help furnace owners evaluate performance and operational impacts before application. Accurate simulation of the enhanced heat transfer from a coating requires spectral radiation calculations which can account for both the absorbing/emitting bands and the transparent windows in the combustion gases. REI's *ADAPT* CFD code was run with spectral calculations to simulate an idealized furnace with and without an enhanced emissivity coating. Changing the wall refractory emissivity from 0.4 to 0.9 resulted in a 1.5% increase in furnace efficiency, a 4°C increase in process fluid exit temperature, and a 29°C drop in gas temperature at the furnace arch. Data from successful field applications of the Emisshield high emissivity coating technology showed fuel savings of 7-8% in a glass melting furnace and 5-6% in several cracking furnaces due to more efficient heat transfer. Production output and run lengths between decoke cycles were also enhanced in several cracking furnace applications. The durability and additional heat conduction behavior of coatings were not addressed in this work.

5. Acknowledgments

CFD graphics were created with Fieldview software by Intelligent Light (http://www.ilight.com). Coating emissivity information and process furnace data were provided by Emisshield, Inc.

6. References

[1] Adams, B., Tang, Q., Ma, J., Brown, D., 2007, "Modeling Combustion in Pyrolysis Furnaces with Next Generation Low NOx Burners," *AFRC-JFRC International Symposium*

[2] Tang, Q., Denison, M., Adams, B., and Brown, D., 2009, "Towards Comprehensive Computational Fluid Dynamics Modeling of Pyrolysis Furnaces With Next Generation Low NOx Burners Using Finite-rate Chemistry," *Proceedings of Combustion Institute, Volume 32, Issue 2, Pages 2649-2657.*

[3] Jamaluddin, A. S. and Smith, P. J., 1988, "Predicting Radiative Transfer in Rectangular Enclosures Using the Discrete Ordinates Method," Combustion Science and Technology, Vol. 59, pp. 321-340.

[4] Denison, M.K., Webb, B.W., 1993a, "A spectral line based weighted-sum-of-gray-gases model for arbitrary RTE solvers," ASME J. Heat Transfer, vol. 115, pp. 1004-1012.

[5] Denison, M.K., Webb, B.W., 1993b, "An absorption-line blackbody distribution function for efficient calculation of total gas radiative transfer". J. Quant. Spectral Rad. Transfer, vol.50, pp.499-510.

[6] Denison, M.K. and Fiveland, W.A., 1997, "A Correlation for the Reordered Wavenumber of the Wide Band Absorptance of Radiating Gases," ASME J. Heat Transfer, Vol. 119, pp. 853-856.

[7] Edwards, D. K., 1976, "Molecular Gas Band Radiation," Advances in Heat Transfer, Vol. 12, pp. 115-193, Academic Press, New York.

[8] Kleeb, T., Fausey, B., 2010, "Fuel Savings with High Emissivity Coatings," 71st Conference on Glass Problems, Oct. 19-20, Columbus, Ohio.

Revamping, Debottlenecking and Optimizing Ethylene Plants using Dual Enhanced Heat Transfer Solutions

Jeremy Provost
Manager Heat Transfer Department
Technip
Paris, France
j.provost@technip.com

Thomas Lang
Manager Technical Marketing – Process Technology
Wieland Thermal Solutions
Graf-Arco-Strasse 36, D-89079 Ulm
thomas.lang@wieland.de

Prepared for Presentation at
American Institute of Chemical Engineers
2015 Spring Meeting
11th Global Congress on Process Safety
Austin, Texas
April 27-30, 2015

UNPUBLISHED

Revamping, Debottlenecking and Optimizing Ethylene Plants using Dual Enhanced Heat Transfer Solutions

Jeremy Provost
Manager Heat Transfer Deptartment
Technip
Paris, France
j.provost@technip.com

Thomas Lang
Manager Technical Marketing – Process Technology
Wieland Thermal Solutions
Graf-Arco-Strasse 36, D-89079 Ulm
thomas.lang@wieland.de

Keywords: enhanced heat transfer, heat exchanger, efficiency improvement, plant optimization, quench-water fouling service

Abstract

Enhanced heat transfer tubes can play a crucial role in the optimization of key heat exchangers both within existing as well as new large scale grassroots ethylene plants. Technip and Wieland have jointly developed efficient enhanced heat exchanger solutions for more compact, cost effective as well as efficient shell and tube heat exchangers especially in the cold-end section of ethylene plants, such as fractionation respectively splitter reboiler and condenser services.

1. Background Ethylene

In the recent years the ethylene business encountered a complete change of its market structure caused by the shale gas boom in North America and the large trend for integration of refining and petrochemical industries, e.g. in Asia. Thus the main drivers for the optimization of existing plants have become capacity upgrade, plant flexibility as well as adaptation to changing feedstock and reduction of carbon foot print. For new plants the minimization of the investment costs as well as carbon foot print are the keys to a successful business. In both cases fully adapted and highly reliable solutions are required to match the lifetime of 20 and more years of plants often located in harsh environment.

2. Enhanced Heat Transfer Technology Background

Low-finned and dual enhanced tubes are widely accepted as standard solutions for several decades in shell and tube heat exchangers in a wide range of industries such as the air-conditioning, power generation, hydrocarbon processing as well as machinery and plant

Note: Do not add page numbers. Do not refer to page numbers when referencing different portions of the paper

technology industry. For the hydrocarbon processing industry, substantial progress has been made in the recent years towards the qualification of enhanced heat transfer technologies for clean and low fouling services in key reboiler and condenser heat exchangers both in refrigeration and thermal distillation processes, e.g. in natural gas liquefaction and ethylene plants.

Over the last 18 years Wieland dual enhanced GEWA-PB and –KS tubes, see Figure 1, have been developed and applied to a standard technology in Liquefied Natural Gas (LNG) and ethylene plants [1, 2, 3]. The primary applications are for clean refrigeration and separation processes with clean C2 and C3 hydrocarbon and refrigeration type fluids. In LNG plants these are chillers and condensers for the propane pre-cooling cycle to cool the feed gas and the mixed refrigerant, respectively. For ethylene plants the preferred application is for distillation towers: reboilers and condensers as well as refrigerant condensers for the ethylene and propylene refrigerant loop.

To further broaden the application of enhanced heat transfer technologies, interesting applications for both GEWA-PB and –KS tubes have been identified for heat exchanger equipment working with cooling water as well as quench water in ethylene plants [4].

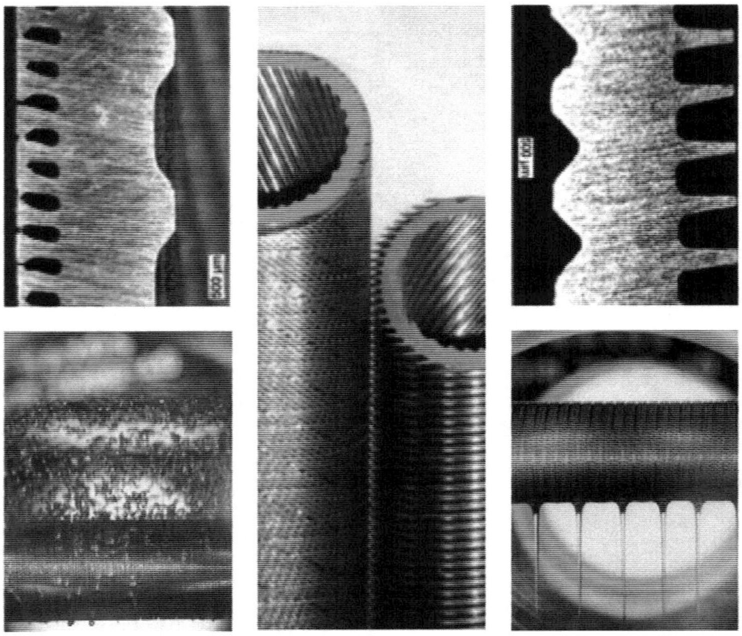

Figure 1: Wieland Dual Enhanced GEWA-PB and –KS Tubes

3. Benefits for Revamp Applications

Dual enhanced heat transfer tubes offer multiple benefits when revamping / debottlenecking an existing plant:

- Capacity improvements of existing shell and tube heat exchangers installed in-place by supporting increased plant through-put

- Minimizing temperature differences and cold approaches allowing for maximizing efficiencies of refrigeration systems
- Improved operating range for critical rotating equipment by minimization of pressure differential between refrigerant reboilers and condensers
- For some particular revamps with new equipment to be provided, size reduction as well as minimization of the number of shells per unit resulting in both plot space and cost reductions
- Improved heat integration using in a more efficient manner waste heat e.g. from quench water

These enhanced heat transfer tubes are dedicated to horizontal shell and tube heat exchangers like thermosiphon and kettle type reboilers as well as overhead condensers. A wide range of pioneering references in revamp and grassroot applications since 2000 confirms the full thermal, mechanical and operational long-term viability of these solutions, see Table 1.

Table 1. Dual enhanced tube solutions available today.

Application	Service	Enhanced Tube
Clean fluids	Depropaniser condenser C2 refrigerant condenser C3 splitter reboiler C3 splitter condenser C2 splitter condenser De-ethaniser condenser Ethane vaporiser C2 splitter reboiler	GEWA-PB
	C3 splitter condenser Propylene refrigerant condenser Propane refrigerant condenser	GEWA-KS
Low fouling fluids	C3 splitter reboiler with quench water from naphtha cracking	GEWA-PB

Different selected case studies will illustrate the benefits of enhanced heat transfer tubes depending on the individual application situation:

- Reboiler for a de-ethaniser overhead condenser service
- C3 refrigerant condenser performance improvement
- C3 splitter reboiler service in an open heat pump

Further to this, a recent development with an enhanced C3 quench-water heated reboiler did show very attractive and stable operation [4]. Data collected and results from a C3 stripper unit installed at a naphtha-based steam cracker plant of a European ethylene producer will be presented. Two case studies will demonstrate the suitability and robustness of this technology in quench-water fouling services:

- C3 stripper reboiler with quench-water heating for heat integration

- Enhanced air-cooler for quench-water cooling as a most compact design.

All presented case studies rely on tubes with optimized heat transfer thanks to structured surfaces on the tube out- and inside. The outside enhancements are taylor-made either for enhanced nucleate boiling, enhanced condensation and gas heating or cooling. On the tube inside helical fin geometries are adjusted to support optimized heat transfer for 1-phase liquid and gas or 2-phase flow. The design correlations and performance were verified in laboratory and scale-up testing as well as field qualifications.

To achieve this level of performances and, more important, to maintain it over time, the enhancement structures are manufactured through a cold rolling process used for the classical integrally low fin tubes by plastically forming fins from the original material out of the plain tube wall. The tubes discussed here are available in various grades of carbon steel (standard, low temperature carbon steel, 3.5 % Ni alloy) allowing for a wide range of operating temperatures, including cryogenic conditions.

4. Case Studies

The objective is to evaluate the interest to simply change the tube bundle or the shell of an existing heat exchanger, with dual enhanced tubes in case of debottlenecking or retubing, by leaving plot size, nozzles and piping unchanged. Five services from recent projects of Technip have been selected for closer review.

4.1. De-Ethaniser Condenser

The first case is the de-ethaniser condenser, in which the overhead stream of the de-ethaniser column is condensed on the tubeside by vaporizing a C3 refrigerant stream on the shellside. The existing heat exchanger is a kettle K TEMA type equipped with 30 foot long and ¾" OD plain tubes. By switching to GEWA-PB tube with the same number of tubes, the overdesign of this exchanger reaches 100% but the tubeside pressure drop exceeds the allowable pressure drop.

Using 1" OD tubes reduces the available surface area and increases the free flow section tubeside which leads to 50% overdesign for the expected allowable pressure drop.

Two design alternatives are described in Table 2 giving an idea of how 50% overdesign could be used.

In the tabulation are summarized the propane saturation temperature, the duty of the existing heat exchanger and the values that can be reached when using GEWA-PB tube in 1" OD.

Table 2. De-Ethaniser Condenser - Design Comparison for Minimal LMTD (1) and Maximal Capacity (2)

		Original	GEWA-PB 1"OD	
			1	2
C3 saturation temperature	°C *(°F)*	-17.6 *(0.32)*	-14.6 *(5.72)*	-
Heat exchanged	MW *(MMBTU/h)*	6.2 *(21.14)*	-	12.7 *(43.32)*
Delta		-	3°C *(5.4°F)*	+104%

Calculation 1: By keeping the original process conditions, 50% overdesign allow to decrease the propane boiling saturated temperature by 3 K (which means a pressure reduction of around 0.3 bar (4.5 psi) which is very interesting in the case of propane compressor power limitation.

Calculation 2: Keeping the original pressures and temperatures, and increasing the flow rates, 50% overdesign allow to double the duty exchanged which is very interesting in case of increase of production.

In a real case the solution will probably be a combination of those two theoretical calculations, taking into account the specific plant characteristics and objectives.

4.2. C3 Refrigerant Condenser

The second service studied is a propylene refrigerant condenser in which propylene refrigerant from the fourth propylene refrigerant compressor stage is fully condensed on the shellside, using closed loop cooling water on tubeside.

The existing heat exchanger unit consists of 6 X TEMA type shells installed in parallel and equipped with 40 foot long 1" OD low fin tubes.

By switching from low-fin to GEWA-KS tube with the same number of tubes, the overdesign of this exchanger is 25% using only 50% of the allowable tubeside pressure drop. Using 3/4" OD tubes increases the available surface area leading to 73% overdesign for the expected allowable pressure drop.

In the tabulation, Table 3, are summarized the cooling water outlet temperature and flowrate, the internal temperature pinch and the duty of the existing heat exchanger:

Table 3. Propylene Refrigerant Condenser - Design Comparison for Maximal Flow Rate (1) and Minimal Temperature Pinch (2)

		Original	GEWA-KS 3/4" OD	
			1	2
$T_{out\ CW}$	°C *(°F)*	43 *(109.4)*	45 *(113)*	-
CW Flowrate	T/h *(1 000 lb/h)*	28 200 *(62 170)*	31 150 *(68 660)*	-
Internal temperature pinch	K *(°F)*	6 *(11)*	-	3.5 *(6.3)*
Heat exchanged	MW *(MMBTU/h)*	164 *(559)*	252 *(864)*	-
Shell number	#	6	-	-
Delta			+54%	2.5 K *(4.7°F)*

Calculation 1: Here the goal with GEWA-KS tubes is to increase the duty. By increasing the cooling water flowrate up to the maximal allowable pressure drop and cooling water outlet temperature, the available overdesign is fully used. Therefore by switching from 1" OD low fin tube to ¾" OD GEWA-KS, we are able to increase the duty by more than 50%.

Calculation 2. By keeping the original process conditions, and decreasing the internal temperature pinch 73% overdesign allow to decrease the C3 refrigerant saturation temperature and pressure. In contrary to the previous case for this heat exchanger it makes more sense to look at the internal temperature pinch than only at the absolute C3 refrigerant saturation temperature.

Nevertheless, the conclusion is the same for both services: the reduction of the saturation pressure in some case could be of an interest in case of compressor power limitation.

As an option for a 3rd optimization, the reduction of the number of shells can be proposed from 6 in the original design 4 with the GEWA-KS tube.

In the case of critical cooling water qualities (e.g. open loop with cooling water coming from cooling towers), an alternative solution is still available thanks to GEWA-K low fin tubes in duplex material. The benefits will not be as important as with GEWA-KS dual enhanced tubes but it will improve the performance thanks to an increase by a factor 2.5 of the available heat transfer area for condensation on the shell side. For existing equipment in carbon steel duplex material can help very efficiently to increase exchanger lifetime and minimize maintenance cost.

4.3. C3 Splitter Reboiler (Open Heat Pump)

The next service studied is a C3 splitter reboiler operating as an open heat pump.

The bottom of the C3 splitter is vaporized on the shellside in a thermosiphon mode by condensing tubeside the overhead of the C3 splitter after this stream went through the C3 splitter overhead compressor.

The existing heat exchanger unit consists of 4 X TEMA type shells operating in parallel equipped with 40 foot long ¾" OD plain tube. By switching from plain to GEWA-PB tubes with the same number of tube, the overdesign of this exchanger is 50% for a tubeside pressure drop below the allowable one. The resulting opportunities are summarized in Table 3.

Table 3. C3 Splitter Thermosiphon Reboiler - Design Comparison for Maximal Duty (1) and Minimal Temperature Pinch (2)

		Original	GEWA-PB 3/4" OD	
			1	2
Cold temperature approach	K *(°F)*	7 *(13.2)*	-	5 *(9.3)*
Heat exchanged	MW *(MMBTU/h)*	83.33 *(284.5)*	121 *(413.8)*	-
Delta		-	+45%	2 K *(3.9 °F)*

Calculation 1. The C3 splitter overhead flow rate is increased and the 50% overdesign allow to increase the duty by 45% which is very interesting in case of increasing the production. The tubeside pressure drop in this case is still lower than the allowable one.

Calculation 2. In this case it is very interesting to look at the cold end pinch. As in the previous case this reflects the saturated temperature modification. The 50% overdesign this time can be used in a saturation temperature reduction of 2°C.

4.4. C3 Stripper Reboiler (Heated by Quench Water)

Quench water is used to cool and clean the cracker gas stream. In naphtha based crackers two quench systems with oil and water are used, see Figure 2. The quench water typically leaves the quenching tower between 80 to 90 °C and has to be cooled down either by heat integration, here the C3 stripper reboiler is the most prominent service and / or by air and water based coolers.

Figure 2: Schematic of a Quenching System for Naphtha based Crackers

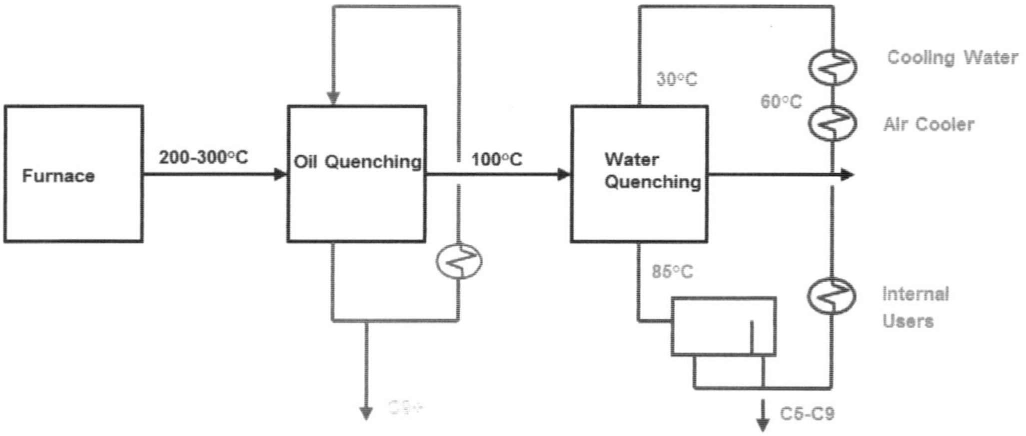

Technip and Wieland did collect data on a horizontal C3 stripper reboiler shell and tube heat exchanger equipped with GEWA-PB tube with quench water flowing on the tubeside , it allows to propose a dedicated low fouling internal tube geometry for such quench water applications when quench water is issued from a naphtha cracking unit [4]. The record of collected fouling data over more than two years allows for a conclusion of a very limited and stable fouling level. Compared to typical design fouling coefficients in the range of 0.0003 to 0.0006 m^2.K/W the measured fouling coefficient mainly below 0.0001 (0.00057 ft^2°F h/Btu) with maximal level up to 0.00015 m^2·K/W (0.00085 ft^2°F h/Btu) is by a factor of three smaller. This new revised level of a fouling coefficient for good design with the GEWA-PB tube does substantially improve the heat exchanger design optimization capabilities. Already the size comparison between plain and GEWA-PB tubes of the stab-in reboiler does show the substantial potential involved, see Table 5.

Table 5. Design Comparison of Stab-in Reboiler for C3 Stripper Design for Quench Water Field Qualification

		Plain Tube (existing)	**Double enhanced GEWA-PB tube (new)**
Shell internal diameter	mm	662	
TEMA heat exchanger type		B-U	
Number of U tubes		153	82
Tube outside diameter	mm	25.00	19.05
Number of passes		4	6
Effective area (based on outside bare area)	m^2	24.4	9.1
Tube side quench water velocity	m/s	0.31	1.50
Internal fouling thermal resistance	$m^2.K/W$	0.00026	0.00026
External fouling thermal resistance	$m^2.K/W$	0.00017	0
Total fouling thermal resistance	%	Over 40%	Over 65%

The primary criterion when designing the C3 stripper reboiler with dual enhanced GEWA-PB tube has been to maximize the tubeside velocity (1 m/s min. as guideline) up to the allowable pressure drop limit given by the pump. Even with a conservative tubeside fouling coefficient of 0.00026 $m^2 \cdot K/W$ the design comparison at constant duty does show a reduction of the number of U-tubes by factor 2. During the qualification the duty of the enhanced bundle has been even larger when the overall fouling value measured resulted in the range of 0.0001 $m^2 \cdot K/W$.

4.5. Quench Water Air Cooler

The low fouling internal structure used in the previous study and incorporated in a design with Dual Internally and Externally Structured Tubes for Air-coolers, called DIESTA, Figure 3, can be used to improve performance of air cooled quench water services when quench water comes from a naphtha cracking unit. DIESTA technology has been recently developed by three industry partners Technip, Wieland and GEA-BTT manufacturer of air coolers.

Figure 3: Dual Internally and Externally Enhanced Structured Tube for Air-Cooler (DIESTA)

The potential of DIESTA has been reviewed in a previous air cooler where quench water is cooled from 80 °C to 57 °C, Conventional air cooler design is resulting in 16 bays with 50 foot long tubes and 60-hp motors. Using DIESTA the same heat duty of 116.5 MW can be extracted using 14 bays. Given the overall cost of air cooler units including plot layout, steel structure and piping this results in a very interesting solution.

In addition such solutions can be applied very attractively in revamp and debottlenecking situations with existing plants where both space and mechanical loads of the structure are limiting other alternatives. For example, based on a typical design, the same quench water air cooler (16 bays) has been studied with DIESTA solution for several revamp configurations (see Table 6).

Table 6. Quench water air cooler - Design Comparison for Maximal Duty (1), Minimal Temperature Pinch (2) and higher summer air temperature (3)

		Original	**DIESTA**		
			1	**2**	**3**
Quench water inlet / outlet temperatures	°C *(°F)*	80 / 57.3 *(176/135.1)*	80 / 57.3 *(176/135.1)*	80 / 54.5 *(176/130.1)*	80 / 57.3 *(176/135.1)*
Heat exchanged	MW *(MMBTU/h)*	116.5 *(398)*	139.7 *(477)*	132.5 *(452)*	116.5 *(398)*
Air side temperature	°C *(°F)*	42 *(107.6)*	42 *(107.6)*	42 *(107.6)*	46.5 *(115.7)*
Tube pass pattern		Even	Even	Even	Even

First of all same horsepower and same allowable pressure drop are considered during revamp. Furthermore no piping modification is foreseen as well as no steel structure as the weight of a DIESTA tube is similar to the standard tube.

Calculation 1: The available overdesign thanks to the use of DIESTA allows to cool a quench water flow rate increased by 18%.

Calculation 2: The available overdesign thanks to the use of DIESTA allows to get a colder quench water at outlet of the air cooler by -2.8 °C (-5°F) for the same quench water flowrate.

Calculation 3: The available overdesign thanks to the use of DIESTA allows to keep cooling capacity in summer season when air ambient temperature exceeds the design air temperature by 4.5 °C (8.1°F).

So whatever is the issue when revamping or debottlenecking, DIESTA can be a very helpful solution to increase the performance of the quench water air cooler.

5. Conclusions

Low-finned and dual enhanced tubes are widely accepted as robust and proven solutions in term of thermal design and mechanical integrity since the 70's. In this paper, we have presented in multiple cases the optimization potential of enhanced heat transfer designs in revamp and debottlenecking situations with existing ethylene plants. Very attractive and beneficial opportunities exist depending on the plant situation towards efficiency or capacity gain or improvements on maintenance as well as operability of a plant.

Most of the heat exchanger operation improvements can be accessed very easily and typical realization can be planned and carried through during normal or partial plant shut down.

The available technologies thereby fit very nicely in the today market demand to support the expansion of petrochemical plants due to the shale gas boom in the US as well as integration of existing refining and petrochemical industries in Asia.

6. References

[1] Brigitte Ploix, Thomas Lang, Dual Enhanced Tubes for the Hydrocarbon Processing Industry – From Debottlenecking to Grassroots, LNG 13, Oran, 2010

[2] Brigitte Ploix, Thomas Lang, Application of Dual Enhanced Tubes for Ethylene and LNG Services, AICHE Spring Meeting, San Antonio, 21-25 March 2010

[3] Brigitte Ploix, Dominique Gadelle, Rodolphe Peronnet, Increase LNG Capacity by Using Dual Enhanced Heat Transfer Solutions, LNG17, Houston USA, 2013

[4] Jeremy Provost, Jean El-Hajal, Thomas Lang, Improve Heat Transfer in Quench Water Services, Hydrocarbon Processing Journal, pp. 53-59, June 2014

401600 Heat Transfer Improvement in the Concentrator System

Abstract:

In study we analyze, model and develop solutions to improve the performance of heat exchangers that are a part the larger system used to increase solution concentration. Concentration is done via pressurizing and heating of solution followed by the drop of pressure to induce vapor generation (flash) in a separating vessel. Heating of solution is done in two exchangers connected in series which needed performance improvement to support plant production increase. We present the analysis of the exchanger system, methodology for its modeling in Aspen and its validation using plant data. Specific solutions leading to increased heat duty in exchanger are also discussed.

Why are we still controlling process plants like we did in the 1950s?

David Huffman – ABB Inc.
Presented at: AIChE Spring Meeting 2015, Austin, TX
Topical 3: Manufacturing for the 21st Century
Session: Reliability, Safety and Control for Chemical Manufacturing

Abstract

The early days of process automation had operators controlling individual loops and devices from field control boxes. Operators today still control the same way, but now they sit in a chair in front of a computer workstation. In many ways, the only thing that has changed is the location and the number of loops one person can manage. Controllers have increased in power and capabilities but few companies have changed their automation to take advantage of these improvements. State Based Control is a way to make significant improvements in production KPIs and plant safety. Methodology is consistent with ISA106.

Introduction: A walk down memory lane

In the early days of process automation, pneumatic controllers were the gold standard. These controllers were located near the control element. For those that are too young to recall these devices, they were individual controllers contained in boxes mounted on a wall, post, or sometimes directly on the process equipment. The operators had to walk from place to place and open the boxes to make adjustments to the control setpoints. Circular charts were sometime included to allow the operator to see the history of the controlled value.

As the technology progressed and the number of control points increased, having to move extensively throughout the physical process became a challenge so pneumatic controls moved inside to a control room, often distributed across large process graphic boards to allow the board operator to visualize where the control point was in the process. Circular charts were replaced with strip charts to record the history of the data.

Gradually, electronic single loop controls began to replace pneumatics, but the method of controlling did not change.

In the late 1970s, Distributed Control System (DCS) technology began replacing single loop controllers. The number of control loops an operator could deal with increased, but operators still generally interfaced with individual device and controller faceplates to run the process.

DCS technology has evolved significantly in the last 40 years especially when considering the graphical interfaces available today for the operators to visualize their scope of control. But in addition to the graphical interfaces, the controllers have become significantly more powerful and incorporating many types of modern control technics including advanced model based control and fuzzy logic.

Each of the evolutionary changes that have occurred in automation has allowed a single individual to control more and more points. But in most process plants today, the operator is still managing their scope of control one point at a time. Instead of standing in front of pneumatic controller box, the operator is sitting in front of a computer monitor clicking on graphics and opening a faceplate. An operator can have a much larger scope of control, but that control is still happening in much the same way it did 60+ years ago.

Compare this to technology in our everyday lives

Think about processes we deal live within our everyday lives that have changed and greatly improved. Consider the following three examples.

Parallel parking your car has gone from a manual "hit or miss" process to one that, in many vehicles today, is a fully automated process. When I learned to drive, parallel parking was part of the test and was frequently needed and required skills in selecting a spot I could fit my car in to, my ability to properly steer the car into the parking place, get it a reasonable distance from the curb, while not hitting either of the cars I was parking between. Today, the technology is evolving in cars to where the vehicle will tell you if the parking spot is in fact large enough to fit the car into and then will automatically park itself without human interaction. The process of parallel parking a car is now much safer and faster. And individuals with low skills or no skills to actually parallel park a car can now achieve the necessary results because of the available technology.

Avoidance speed control is another evolving technology adding safety to driving. When I started driving, cars didn't have cruise control to aid in maintaining a safe, or at least desired, speed on the open road. A few years later, the technology was available, but it was useless in traffic or on city streets. Today avoidance speed control can not only maintain a desired speed, but in some car models it can slow the vehicle as it approaches the vehicle ahead and in many now, will actually stop the car. This combination of features is especially important as we find that many drivers are more and more distracted by technologies that are both part of the car or portable. This same technology will return the car to the desired speed when conditions allow automatically without the need for the driver to take any actions other than steering. Some car manufactures claim the technology can actually be used in stop and go traffic. The operator still has to steer the car, but maintaining speed, slowing, stopping and or resuming setpoint speed conditions can be handled by the controls built into the vehicle.

My last example is my coffee pot. The technology has evolved from the percolator that was in kitchen during my childhood, to the devices that you could fill the night before and have the coffee ready when I woke up, to today's systems that can not only make coffee, but many other hot beverages also. Flexibility has increased significantly from where one kind of coffee could be brewed in large quantities to now each individual can have their favorites. The older technologies generally made one pot of coffee and the user then poured into a cup size of their choosing and frequently a lot of the final product was wasted. Today, the coffeemaker uses a small quantity of coffee resources and the user can select a size that not only is what they want to drink, but is also brewed consistently regardless of the serving size. And many devices can also deliver hot water in measured quantities to use with beverages that may not come in containers that fit into the brewing system. Small quantities of water are heated at one time reducing energy usage and reducing water resource waste when larger pots were brewed and then not fully consumed.

Automation in everyday processes is increasing convenience, providing consistent quality, allowing less skilled individuals to perform at expert levels, reducing resource waste, and making activities safer. This list should sound familiar as it is generally parallels the same performance requirements we find in most corporate business statements.

So how can our process automation enable and contribute to these performance criteria for your process?

State Based Control can be a key contributor

State Based Control, or sequential control, can offer continuous process owners improvements in key performance indicators that include those presented above. Businesses dealing in batch manufacturing figured this out twenty or more years ago. Batch process operators were manufacturing products, following written procedures but operating the process controls manually. As control system functionality increased, sequential controls evolved to automate most if not all of the processing procedures within the DCS.

You may be wondering how sequences apply to continuous process operations. But procedures are applied to continuous processes almost every minute of every day. Besides, continuous processes are not truly continuous. They are built and started for the first time. They operate for a period of time (months or years) and eventually shutdown for turnarounds or maintenance. And then the process of startup, operate and shutdown are repeated, perhaps many times in the lifecycle of a facility. Each of these phases has its own special set of procedures.

Given this lifecycle, it is probably easy to understand that procedures apply to the startup and shutdown phases of the lifecycle. But what about the operational phases of the lifecycle? Most plants have a large quantity of procedures that were written when the plant was first built to operate it. These procedures outline, at least in a generalized way, how the plant is to be operated to make prime product. Many times procedures illustrate how to deal with expected abnormal situations during operation. As the process is operated over time those procedures are modified, improved, and adapted to product and processing modifications. And if the process manufactures varying grades or products, there are always transitions between those that require changes to operating conditions that usually have some form of procedural steps to make the changes. And then there are other continuous processes that have equipment that is taken in and out of service during normal operation to regenerate catalysts or cycle the equipment for other reasons. All of these instances rely on knowledge of procedures to maintain product quality and or safe operating conditions.

State Based Control can provide benefits in five key topical areas:

- Productivity increases from situational optimization (making the best of a bad situation)
- Improve asset utilization – both physical plant assets and human assets
- Improve responses to abnormal process conditions

- Creates a platform for best practices knowledge capture to improve everyone's performance
- Improve the process safety contributions of the basic process control system (BPCS)

Productivity increases are available because automated procedures can generally manipulate the process faster than people can. Startups can happen faster since when step triggers are reached the procedure will move on. They do not wait for the operators to move about the plant, shift changes, meal breaks or other situations that draw an operator's attention away from the actions. A faster process startup means more full-rate prime production. If process transitions are required, automated procedures can frequently achieve them faster and more accurately. This minimizes transitional material and provides more usable product. Shorter transitions also means more time not spent in transitions and making prime product over the course of each production year. In abnormal situations, automated procedures may be able to transition or park the plant into a suboptimal, but safe-running operating state. Doing so may prevent an unnecessary shutdown, not only maintaining production, but also avoiding the usually more dangerous personnel exposure situations of a shutdown and startup.

The increased production periods identified above result in increased physical asset utilization. Other opportunities for asset improvement can come from automated procedures used to start and stop equipment with properly ordered step every time. One example is starting pumps. Done improperly, the pump can blow a seal or experience other damage resulting in repairs and possibly downtime. But automated procedures also allow for increasing personnel utilization. Automating procedures removes the need for the "constant contact" a board operator is usually required to perform in managing equipment through hundreds of faceplates. Process observation is still required, but it can be applied to maintaining Key Performance Indicators (KPIs) or seeking to further improve on the operating conditions rather than just operating. Automated procedures also mean that all operators perform at the same best practices level. Operators with limited board experience can still manufacture during normal production periods and through abnormal conditions handled by the procedures at the same performance levels as much more experienced operators. Additionally, there are techniques related to State Based Control definitions that act as evergreen documentation. This significantly increases the effectiveness of the engineering staff responsible for maintaining the automation system over time. Documentation is always current and accurate so that when changes, upgrades, or expansions are required, the staff does not have to go back and make sure the current information is accurate. Additionally, depending on the system some of these tools can act as tools that can be used to directly translate the definitions directly into configurations for the controllers removing large quantities of hours normally required to translate a definition into downloadable controller code.

Responses to identifiable abnormal situations can be managed rapidly, accurately, and safely. For example, a distillation tower that relies on processing a product stream through a three-bed dryer unit (primary bed, secondary bed, regenerating bed) will require adjustments if the regenerated bed cannot be returned to service when the primary bed saturates. In plants without an automated procedure to handle this situation, the operator has dozens of adjustments to make to try and position the operating conditions of the tower to tolerate continuing operation on only one dryer bed, or about half rates. Depending on the skills of the operator, they may not be able to achieve the proper conditions and either need to move the tower into total reflux or even take the whole unit into a standby state until the a second dryer bed is available. Alternatively, automated procedures provided with a trigger key from the process or from the operator, can make the necessary adjustments and keep the tower running at the reduced rate conditions. When normal operation can be reset and when the operator can give attention to the situation, they again need to manipulate a large group of faceplates to return the unit to full operation. If automated procedures are in place, the reset trigger will automatically return the unit to full service while the operator can focus on potentially other important requirements.

During the abnormal period, a typical plant would experience a high number of alarms. The functionality of State Based Control provides for hiding unimportant alarms from the operator. As equipment is activated, deactivated, or conditions change that would normally result in alarms, the procedures can activate and deactivate alarms appropriately and change alarm trip points for new conditions as warranted. This can remove a great deal of the environmental interference that operators usually need to consider during an abnormal condition creating a much improved and possibly safer operating environment.

The fourth benefit of this discussion is knowledge capture. Operating teams rarely perform daily functions by reading from procedures. The complex operations of a plant are learned and committed to memory during training. Each person, however, remembers things slightly differently, especially over time. It is not uncommon to find that a group of experienced operators will often perform a given procedure with slight differences from one to the other when working from memory. None are wrong, and none may be completely correct. However, when reviewed among the operators, usually a single best practice can be derived from the combinations. It is these best practices that are appropriately captured into automated procedures for everyone to use. It is also beneficial for procedure infrequently used, like startup and shutdowns, to be committed to a mostly automated condition. In many facilities today, the timeframe between turnarounds is generally being extended, in some case to six or more years. Over such long timeframes, it is not uncommon to have a large number of the operations staff that have never been involved in a shutdown/startup situation. Having automated procedures can maintain a high degree of accuracy and safety in the execution of the procedures even if there is low degree of experience in the people. This condition is likely to worsen over the next few years as a large number of the experienced people in our facilities retire. Capturing the process operating knowledge of these experienced people will be paramount for the future success of the businesses they support today.

Safety is always a key concern in process operations. When process owners evaluate layers of protection, especially when trying to determine the safety integrity level (SIL) required for safety instrumented functions (SIFs) there is a consideration for the contribution of the BPCS to provide safety before a dangerous situation is reached. Basic control of critical parameters and alarm management are two of the criteria that get consideration in these analyses. In traditional control designs, these focus on aiding the operator to try and prevent an event from reaching the point where a safety system takes over and parks the plant into a safe state. But with State Based Control both the basic control and alarm management can be greatly improved to further enhance the operator's ability to make critical decisions. The same concepts discussed above about abnormal situations apply to the safety consideration. Faster responses, perhaps selecting conditions difficult for operators to achieve on their own, and the ability to manage a significantly larger number of loops and controls simultaneously provide opportunities to prevent reaching a safety system trigger point. State Based Control alarm management is also a key contributor to process safety. Using the sequence steps to proactively control alarm enabled and disabled states and alarm setpoints is critical to truly effect alarm management as identified within the directives of ISA18.2.

In addition to the above safety concerns, the increased automation that normally accompanies a State Based Control implementation can also contribute to general process safety and preventing injury or hazard exposure for operators. Typically many more valves will be automated removing the need for operators to manually open or close them, removing strain opportunities. It is also not uncommon to find that the increased automation removes the need to expose operators to heights or areas of the process where exposure to extremely hazardous chemicals are in close proximity as with traditional control designs.

Moving away from 1950s controls

Continuous process owners can now move confidently away from traditional control designs and into a new realm of sequential operations to make measureable improvements in their processes. There have been several progressive companies practicing the techniques for several years now and the benefits are well proven in each one.

Batch manufactures discovered this improvement about two decades ago and have been improving on the techniques ever since. But as batch manufactures started using sequence controls, they discovered that they needed a common language to communicate across their varied industries and to communicate needs with automation vendors. The ISA88 standard provided that and much more.

Taking that learning, ISA106 has been started for the continuous industries. The standard is identified as "Procedure Automation for Continuous Process Operations." The committee produced its first Technical Report on Models and Terminology (TR106.00.01) in 2013 and is close to releasing its second report on work processes related to executing and maintaining State Based Control designs. Efforts are also currently beginning to move TR106.00.01 into a standard format. ISA members can view TR106.00.01 online for free or it is available for purchase by members.

At this time, State Based Control is being applied primarily in the chemicals industry segment by a number of progressive companies. In many cases, it is found in companies that have batch business components and adopted the procedural control from the batch segments to the continuous operations. But State Based Control does not need to be limited to chemicals. There are applications for the techniques in almost any continuous control situation including but not limited to petroleum refining and gas processing, power generation, and pulp and paper.

State Based Control does not have to be a large initial undertaking. It can be applied slowly across your businesses. As described within TR106.00.01, the process should be evaluated to identify units and equipment modules within the unit structures. It is relatively easy to start with the smaller equipment modules, applying procedures to these smaller entities before trying to expand into complete units and then coordinating units as some of the long time users have done.

It is time for continuous process automation to move into the 21st century. Much of the technology we use for everyday activities has more applied automation than our process plant do. The choice should be easy, if based only on the benefits related to process safety.

401752 Building of Safety Knowledge Base for Automatic Synthesis and Validation of the SIS Design for Toxic Gas Facilities

Abstract:

As the use of toxic chemicals is being increased for the manufacturing of high value-added products, the number and loss from related acciedents are being increased globally. The proper design of safety-instrumented system (SIS) is a recommended way as a measure to prevent and safely handle accidents related to hazadous chemicals. However, the design and validation of the SIS satisfying the required level of risk is not an easy task and requires lots of time and effort of experts. This research proposes an organized safety KB and automatic synthesis system for the SIS layout design for toxic-gas manufacturing, handling, and storage facilities. The safety KB encodes the design rules, regulatory codes, and best practices related to toxic gases. The prototype of developed system is validated against chlorine and ammonia cases, and the KB is being expanded to cover the toxic gases for semiconductor manufacturing.

401755 Optimal Selection and Placement of Iot-Based Sensors for Early Detection of Chemical Accidents

Abstract:

In addition to the conventional wired sensors, IoT-based sensors, in a form of fixed as well as mobile, are becoming availble for better monitoring of symptoms and collection of information for early detection and diagnosis of accidents in chemical plants. This reserach introduces an on-going national project of the use of M2M/IoT technology for improved management and response to chemical accidents. System design and simulation-best testing against an industrial setting are also presented.

AUTHOR INDEX

AUTHOR INDEX

American Institute of Chemical Engineers
3 Park Avenue
New York, New York 10016-5991

CURRAN ASSOCIATES INC
proceedings
.com